Urban Sustainability

The Urban Sustainability Book Series is a valuable resource for sustainability and urban-related education and research. It offers an inter-disciplinary platform covering all four areas of practice, policy, education, research, and their nexus. The publications in this series are related to critical areas of sustainability, urban studies, planning, and urban geography.

This book series aims to put together cutting-edge research findings linked to the overarching field of urban sustainability. The scope and nature of the topic are broad and interdisciplinary and bring together various associated disciplines from sustainable development, environmental sciences, urbanism, etc. With many advanced research findings in the field, there is a need to put together various discussions and contributions on specific sustainability fields, covering a good range of topics on sustainable development, sustainable urbanism, and urban sustainability. Despite the broad range of issues, we note the importance of practical and policy-oriented directions, extending the literature and directions and pathways towards achieving urban sustainability.

The series will appeal to urbanists, geographers, planners, engineers, architects, governmental authorities, policymakers, researchers of all levels, and to all of those interested in a wide-ranging overview of urban sustainability and its associated fields. The series includes monographs and edited volumes, covering a range of topics under the urban sustainability topic, which can also be used for teaching materials.

Ali Cheshmehzangi · Jian Zuo · Ayyoob Sharifi ·
Rongpeng Zhang · Abbas Ziafati Bafarasat ·
Jie Zhao
Editors

Designing Healthy Buildings and Communities

Shaping a Climate-Resilient Future

Springer

Editors
Ali Cheshmehzangi
School of Architecture, Design and Planning
The University of Queensland
Brisbane, QLD, Australia

Ayyoob Sharifi
The IDEC Institute
Hiroshima University
Hiroshima, Japan

Abbas Ziafati Bafarasat
School of the Built Environment
Oxford Brookes University
Oxford, UK

Jian Zuo
School of Architecture and Built Environment
The University of Adelaide
Adelaide, Australia

Rongpeng Zhang
School of Architecture and Planning
Hunan University
Changsha, Hunan, China

Jie Zhao
The Weitzman School of Design
University of Pennsylvania
Pennsylvania, USA

ISSN 2731-6483 ISSN 2731-6491 (electronic)
Urban Sustainability
ISBN 978-981-96-3056-1 ISBN 978-981-96-3057-8 (eBook)
https://doi.org/10.1007/978-981-96-3057-8

This Springer imprint is published by the registered company Springer Nature Singapore Pte Ltd.
The registered company address is: 152 Beach Road, #21-01/04 Gateway East, Singapore 189721, Singapore

For health specialists, health experts, and those who genuinely care about the health of our societies and communities worldwide.

Preface

Only when the last tree is cut down, the last fish eaten, and the last stream poisoned, you will realise that you cannot eat money.

—Cree Indian Proverb, One of Canada's largest First Nations

"Healthy Buildings" and "Healthy Communities" have become increasingly important in the face of growing urbanisation and climate change. Designing them rightly to ensure a climate-resilient future is shaped holistically is even more important. Our constructed surroundings (also known as the built environments) have a substantial influence on our general well-being, health, and quality of life; they are not merely a background to our lives. It is more important than ever to reconsider our urban landscapes as cities grow and climate patterns shift. In order to address this need, this book looks at creative approaches that combine the goals of environmental sustainability and human health.

This book is the result of realising that, in the context of climate change, our conventional methods for building design and community development/planning/design are insufficient. The traditional focus on efficiency and economic growth frequently ignores the vital relationship between a community's built environment and health. In addition, our cities need to be resilient and adaptive as new risks are brought about by climate change. It is time for a fundamental change in architectural and community design, one that puts human well-being and climate resilience first. This is something that we could have done much earlier when we brought in the concept of sustainable architecture or sustainable community. While it is not too late, we urge the actors on practice and policy to bring in a much-enhanced health-oriented narrative in how we shape our communities and cities with good quality and healthy buildings.

This book offers a comprehensive examination of how to make this change. It offers real-world examples of how creative ideas are being implemented to create healthier and more resilient urban environments through a carefully chosen array of international case studies. From smaller, building design strategies to addressing issues of more vulnerable communities, the book's case studies cover a wide range of contexts and provide insightful information about the challenges of incorporating health and climate resilience into the built environment. The focus of this volume is on building and community scales, while a follow-up volume would explore the larger urban design and planning examples.

Partially, the book focuses on the successful integration of green building technologies. There is still more potential for green building, even though it has moved from being a niche idea to a commonly used practice. Sustainable materials, energy-saving technologies, and biophilic design features allow green buildings to dramatically reduce their environmental impact while improving occupant well-being. Professionals aiming to promote sustainable design can find inspiration from the case studies in this book, which demonstrate the useful applications of these technologies in diverse settings. Yet, these should not be the limits of what a healthy building or a healthy community could look like. A healthy building or community ought to be more than just a performative design and help improve and transform lifestyles, livelihoods, and ultimately, the well-being of society. Creating this change is essential in how our business-as-usual practices and associated policies could shape the future of our cities and society.

The book examines the social aspects of climate resilience in addition to its technical aspects. The built environments are dynamic social ecosystems where people live, work, and interact. They are not just physical places. The well-being of a city's citizens has a direct impact on its resilience and vitality. Thus, building-level strategies and community-driven initiatives play a critical role in fostering resilience and social cohesion. These programs provide citizens the power to actively shape their surroundings, resulting in cities that are more equitable and inclusive as well as more climate change-resilient. Hence, some of the book's case studies highlight effective examples of community involvement and show how teamwork can produce revolutionary results.

The incorporation of natural solutions into community design development is another key component of the book. Cities frequently invade natural ecosystems as they grow, which lowers biodiversity and raises the risk of climate-related disasters. The same applies when we look at successful examples at smaller scales of micro and meso levels, where buildings and communities could be healthier and promote healthier living and lifestyles. By adding natural elements to the built environment settings of various scales, nature-based solutions offer a means of reversing this trend. These solutions, which range from green walls and roofs to urban forests and wetlands, not only improve the aesthetics of cities but also provide vital ecosystem services like controlling stormwater runoff, air purification, and temperature regulation. The book's case studies demonstrate the effective application of healthy building and community design, providing a manual for communities looking to harness the advantages of nature to promote resilience.

The book explores the wider theoretical and policy frameworks that support the idea of healthy buildings and communities concerning climate-resilient strategies and directions, but it mostly focuses on real-world examples. The built environment and health have a complex and multifaceted relationship that calls for an all-encompassing strategy involving multiple disciplines. This book creates a wide range of perspectives that can influence future research and practice by combining insights from the fields of architecture, architectural engineering (or building services/ science), urban planning, public health, environmental science, and social policy. It highlights how important interdisciplinary cooperation is to reaching resilience and health objectives. The problems caused by climate change are too complex for one discipline or industry to handle alone; instead, a cooperative effort combining a range of resources and skills is needed. This book's case studies demonstrate how these kinds of partnerships can produce creative, practical, and long-lasting solutions.

The lessons learned from the case studies will be invaluable as we navigate the uncertainties of the twenty-first century and show that it is possible to plan and build communities that are both climate-resilient and supportive of human health and well-being. Nonetheless, there is still a great deal of work to be done in order to create built environments that are both resilient to climate change and healthy. The ideas and illustrations offered are meant to act as a springboard for additional research and creativity.

To sum up, this book is more than just a compilation of case studies at micro and meso scales; it is a call to action that asks planners, architects, researchers, and policymakers to reevaluate the planning and construction of cities. We can design urban environments that not only withstand the challenges posed by climate change but also enhance the well-being of everyone who resides, thereby implementing a holistic approach that integrates resilience and health. We believe that this book will encourage a new generation of leaders to take decisive action in the pursuit of healthier, more resilient cities, as it bears witness to the strength of creativity, teamwork, and paradigm-shifting thinking.

Acknowledgements We acknowledge everyone's contributions to this exciting book project. We extend our gratitude to all the authors, co-authors, chapter contributors, and collaborators who have shared their expertise, passion, and creativity with us. This sentiment applies to our earlier work as well, where we deeply value the dedication and vision of all contributors that helped create this book. Such collaborative efforts foster new dialogues, networking opportunities, and a valuable research exchange platform for scholars worldwide.

Brisbane, Australia — Ali Cheshmehzangi
Adelaide, Australia — Jian Zuo
Hiroshima, Japan — Ayyoob Sharifi
Changsha, China — Rongpeng Zhang
Oxford, UK — Abbas Ziafati Bafarasat
Pennsylvania, USA — Jie Zhao

About This Book

"Designing Healthy Buildings and Communities" aims to explore and showcase global case studies focused on creating buildings and communities that promote health while enhancing climate resilience. In an era where climate change increasingly impacts urban environments, there is a critical need for innovative solutions that go beyond conventional practices. This book compiles a diverse range of examples, drawing on real-life projects and potential paradigm shifts that emphasise novel ideas and advanced methodologies.

The featured case studies span various aspects of building and community design, including architectural design directions, technology integration, and technical methods. Each contribution provides practical examples that illustrate the implementation of these concepts in real-world settings. By highlighting innovative approaches, the book challenges the status quo and encourages a shift from business-as-usual scenarios to forward-thinking strategies that prioritise both human health and environmental sustainability.

Key themes include the integration of green building technologies, the adoption of sustainable materials, and the incorporation of nature-based solutions in community design. Additionally, the book delves into community-driven initiatives that foster social cohesion and climate resilience, demonstrating how collaborative efforts can lead to more robust and adaptable cities. Through detailed analyses and comprehensive discussions, this book serves as a vital resource for architects, urban planners, policymakers, and researchers. It provides insights into cutting-edge practices and offers inspiration for future projects aimed at creating a healthier and more sustainable future. By bringing together these global perspectives, the book highlights the importance of innovation and adaptability in the face of climate challenges, ultimately contributing to the development of sustainable and thriving urban landscapes.

Key features and benefits:

- A comprehensive collection of healthy and climate resilience research in the built environment sector.

- A unique collection of scholarly work focused on building-level practices and community-level interventions.
- Offers new perspectives on creating buildings and communities that promote health while enhancing climate resilience.

Contents

Editors and Contributors

About the Editors

Ali Cheshmehzangi is a Professor and Head of the School of Architecture, Design and Planning (ADP) at The University of Queensland, where he leads the School and researches across the fields of architecture, design, and planning/urbanism. He has been in the World's top 2 per cent field leader since 2022, recognised by Stanford University. He is among the top 30 global scholars in the urban sustainability research area. Ali has a wealth of experience, expertise, and a strong commitment to innovation and excellence in the fields of architecture, design, and planning, the three core disciplines of his School at UQ. With a career spanning over two decades, he has made significant contributions to the academic and professional communities, with a focus on sustainable and environmentally conscious design. So far, Ali has published over 500 journal papers, articles, conference papers, book chapters, and reports. He also has 35 other academic books, three of which have received awards at the national, provincial, and municipal levels. He has also received international awards and recognition for his research on urban resilience studies and sustainability research, as well as a Vice-Chancellor's award for his impactful contribution to higher education.

Jian Zuo is a Professor in Sustainable Construction at the School of Architecture and Civil Engineering at the University of Adelaide. His main research interests are related to the sustainable built environment via both technical (e.g. smart technologies) and managerial (e.g. policy intervention and behavioural change) means. In particular, he has undertaken extensive research in construction and demolition waste recycling and management. As the lead researcher, he has been involved in a series of competitive grants in Australia such as Discovery Projects and Linkage Projects funded by the Australian Research Council as well as many other national key projects. He has long been engaged in the study of ecological towns, green buildings, and sustainable development of the construction industry. Published more than 100 papers in top journals, including more than 80 journal papers indexed by SCI and SSCI, Google Scholar indexed more than 28,000, and the H-index is 93. Twelve papers are listed as hot papers/highly cited papers by Incites Essential Science Indicators (ESI).

Ayyoob Sharifi is a Professor at the IDEC Institute, Hiroshima University. He also is a core member of the Network for Education and Research on Peace and Sustainability (NERPS). His research is mainly focused on urban sustainability, resilience, and climate change mitigation and adaptation. He has published extensively on these topics and is ranked among the world's top 1000 influential authors in all fields according to Stanford/Elsevier's Top 2% Scientist Rankings in 2024. Additionally, he received Clarivate's Highly Cited Researcher Award in 2024. Ayyoob actively contributes to global change research programs such as the Future Earth and has served as a lead author for the Sixth Assessment Report (AR6) of the Intergovernmental Panel on Climate Change (IPCC). The goal of his education and research activities is to inform actions towards building sustainable and peaceful communities.

Rongpeng Zhang is a Full Professor and dedicated PhD Supervisor at the School of Architecture and Planning, Hunan University. His primary research pursuits encompass building performance modelling and simulation for intelligent buildings, alongside advanced built environment control for healthy buildings. He was a primary developer of several pivotal modules within the DOE-endorsed EnergyPlus building simulation engine, including VRF-HR, fault detection, and demand response modules. He was a primary developer of Commercial Building Energy Saver (CBES) platform, which obtained the 2019 R&D 100 Award as one of the 100 most significant new technologies in the world. Before joining Hunan University, he led the building science team and conducted a series of research on human-centric built environment control at Well Living Lab, which obtained the 2018 Edison Award. He also served as an assistant professor at Mayo Clinic, and a postdoctoral fellow in the Building Technology and Urban Systems Division at Lawrence Berkeley National Laboratory. He received his Doctorate in Building Performance and Diagnostics from School of Architecture, Carnegie Mellon University and his B.S. and M.S. degrees in Built Environment and Facilities Engineering at Tongji University.

Abbas Ziafati Bafarasat is a Charted Town Planner and Senior Lecturer in Town Planning, School of the Built Environment, Oxford Brookes University, where he also leads the M.Sc. Program in Spatial Planning, and the Certificate in Spatial Planning Studies. Abbas undertook his Ph.D. in Planning and Landscape at the University of Manchester, and his Post Doc at Technical University of Dortmund. His research and professional experience spans strategy and its interdisciplinary applications in enhancing policy and placemaking. Abbas is keen on generative interdisciplinary research and practice systematising knowledge in confronting emerging issues in sustainable development, with some of his outputs including: (i) proposing an inter-discipline of strategic urban design, (ii) introducing behavioural indicators for sustainability monitoring, (iii) proposing the concept of start-it-yourself urbanism, and (iv) conceptualising bottom-up agenda-setting.

Jie Zhao is a Lecturer of architecture in the Weitzman School of Design at the University of Pennsylvania and Head of Delos Labs and an Executive Vice President at Delos. Delos is a science-based wellness real estate company headquartered in New York City. As the creator of the WELL Building Standard, Delos develops programs and solutions that transform our environments into vehicles for health, well-being, performance, and resilience. Jie leads the global research effort for advancing science and innovation. His research focuses on human-building Interactions, including healthy buildings, total building performance, occupant behaviour in buildings, sensing and controls in buildings, health, comfort, and productivity of building occupants, as well as building energy efficiency and sustainability. He has published 20+ academic articles and serves as reviewers for 10+ top tier academic journals and conferences in the building science field. He frequently contributes to public media as a subject matter expert in healthy buildings, such as Wall Street Journal, CBS News, Forbes and Money. Jie received his Ph.D. in Building Performance and Diagnostics from Carnegie Mellon University and Master's and Bachelor's degrees from Tongji University.

Contributors

Ngoran Hassan Afoni Department of Geography, University of Buea, Buea, Cameroon

Sara Aristizabal Delos Living LLC, New York, NY, USA

Ali Cheshmehzangi School of Architecture, Design and Planning (ADP), The University of Queensland, Brisbane, QLD, Australia;
School of Architecture and Civil Engineering, The University of Adelaide, Adelaide, SA, Australia

Francis Deng Clement School of Arts (Architecture), Nantong University, Nantong, China

Seyedeh Saeedeh Hoseinyzadeh Department of Art and Architecture, Payame Noor University, Tehran, Iran

Bilal Hussain Department of Economics, Government College University, Faisalabad, Pakistan

Foka Sherifatu Kinyuy Department of Geography, University of Buea, Buea, Cameroon

Clarius Kuyar Department of Geography and Planning, The University of Bamenda, Bambili, Cameroon

Armin Mehdipour Faculty of Science, Engineering and Technology, The University of Adelaide, Adelaide, SA, Australia

David Mihigo Hangzhou International Innovation Institute of Beihang University (BUAA), Hangzhou, China

Saeedeh Moayedfar Department of Geography, Faculty of Humanities, Meybod University, Meybod, Iran

Syed Asif Ali Naqvi Department of Economics, Government College University, Faisalabad, Pakistan

Ngala Elvis Ngum Department of Geography and Planning, The University of Bamenda, Bambili, Cameroon

Yongka Bernard Nyuyse Department of Geography and Planning, The University of Bamenda, Bambili, Cameroon

Anna Obraztsova Delos Living LLC, New York, NY, USA

Mousa Pazhuhan Geography and Urban Planning, Independent resilience researcher, Tehran, Iran

Ke Peng School of Architecture and Planning, Hunan University, Changsha, P.R. China

Zhimou Peng School of Architecture and Planning, Hunan University, Changsha, P.R. China

Ayyoob Sharifi IDEC Institute and Network for Education and Research on Peace and Sustainability (NERPS), Hiroshima University, Hiroshima, Japan

Ngoran Larissa Tomla Department of Geography and Planning, The University of Bamenda, Bambili, Cameroon

Suiven John Paul Tume Department of Geography and Planning, The University of Bamenda, Bambili, Cameroon

Atiq ur Rehman Department of Economics, Government College University, Faisalabad, Pakistan

Regina Vaicekonyte Delos Living LLC, New York, NY, USA

Rongpeng Zhang School of Architecture and Planning, Hunan University, Changsha, P.R. China

Jie Zhao The Weitzman School of Design, The University of Pennsylvania, Philadelphia, USA;
Delos Labs, Delos, New York, NY, USA

Shutian Zhou School of Arts (Architecture), Nantong University, Nantong, China

Abbas Ziafati Bafarasat School of the Built Environment, Oxford Brookes University, Oxford, UK

Jian Zuo School of Architecture and Civil Engineering, The University of Adelaide, Adelaide, SA, Australia

Foundations of Health and Resilience: Designing a Purpose-Built Sustainable Future by Breaking Business-as-Usual Practices

Ali Cheshmehzangi, Jian Zuo, Ayyoob Sharifi, Rongpeng Zhang, Abbas Ziafati Bafarasat, and Jie Zhao

Abstract Urban resilience and public health are seriously threatened by the growing frequency and intensity of climate change's effects. The interrelated problems that communities around the world face, such as extreme weather events, rising temperatures, deteriorating environmental stressors, and pandemics highlight the urgent need for health-resilient design thinking, strategies, and practices. These occurrences endanger millions of people's lives in addition to infrastructure, especially in crowded urban areas. This introductory chapter examines the health hazards that come with urbanization, including air pollution and mental health conditions that are made worse by the effects of climate change. Additionally, it presents the idea of climate-resilient communities, which actively plan for and adjust to shifting environmental conditions while advancing environmental sustainability and human health. Cities and structures can be designed to improve well-being and withstand the increasing threats of climate change by incorporating health into the framework of climate resilience. The chapter lays the groundwork for practical solutions and promotes a

A. Cheshmehzangi (✉)
School of Architecture, Design and Planning (ADP), The University of Queensland, Brisbane, QLD, Australia
e-mail: A.Chesh@uq.edu.au

A. Cheshmehzangi · J. Zuo
School of Architecture and Civil Engineering, The University of Adelaide, Adelaide, SA, Australia

A. Sharifi
IDEC Institute and Network for Education and Research on Peace and Sustainability (NERPS), Hiroshima University, Hiroshima, Japan

R. Zhang
School of Architecture and Planning, Hunan University, Changsha, China

A. Ziafati Bafarasat
School of the Built Environment, Oxford Brookes University, Oxford, UK

J. Zhao
The Weitzman School of Design, The University of Pennsylvania, Philadelphia, USA

Delos Labs, Delos, New York, NY, USA

A. Cheshmehzangi et al. (eds.), *Designing Healthy Buildings and Communities*, Urban Sustainability, https://doi.org/10.1007/978-981-96-3057-8_1

paradigm change in urban environment design that advances both community health and sustainability.

Keywords Health-resilient design · Climate change · Climate-resilience · Environmental stressors · Sustainability · Healthy buildings · Healthy communities

1 The Urgency of Health-Resilient Design in a Changing Climate

The effects of climate change are becoming more frequent and severe; they are no longer merely hypothetical [1–7]. Millions of people's health and well-being are at risk due to extreme weather events, rising temperatures, and changing environmental conditions that are occurring in communities all over the world. It is now obvious that our built environments—our homes, workplaces, public areas, and cities—must be made to withstand these difficulties as the world deals with this growing crisis [8]. The urgency of this issue, however, transcends merely addressing environmental sustainability [9, 10], resilience and health must be integrated holistically [11–13] into the framework of building and urban design.

Ecosystems, economies, and human lives are all being impacted by climate change on an unprecedented scale [14]. Since 2007, over half of the world's population lives in cities, making them especially vulnerable to the consequences of climate change [15]. Vulnerabilities are particularly evident in developing countries where the majority of the future urban population growth is expected to occur. Urban areas are already being impacted by rising sea levels, heat waves, more precipitation, and natural hazards such as hurricanes and floods, frequently with disastrous results [16]. These occurrences put a strain on public services, disturb infrastructure, and force people to relocate, emphasizing the urgent need for resilience in urban planning and design [17].

Cities around the world are ill-equipped to handle these swift environmental shifts. For example, cities with inadequate drainage systems struggle to control flooding brought on by more frequent and powerful storms, which damages infrastructure, contaminates water, and uproots vulnerable populations. In a similar vein, residents of crowded, inadequately ventilated, and under-resourced neighbourhoods are disproportionately affected by heat waves [18], which are worsened by the urban heat island effect (UHIE) and lead to heat-related illnesses and fatalities. The effects are not equally distributed, with lower-income and marginalized communities suffering the most.

Similarly, a large number of buildings are unprepared or underprepared to deal with these changes because they were built for a climate that no longer exists. Buildings that have inadequate insulation, no natural ventilation, and rely on energy-intensive cooling systems are unhealthy and inefficient because they retain heat, moisture, and pollutants that can worsen respiratory disorders and other health issues

[19, 20]. The need for built environments that can respond, adapt, and shield their occupants from these new threats is growing as climate change picks up speed.

1.1 Rising Health Concerns Linked to Urbanization and Environmental Stressors

There is growing concern about the health risks associated with urbanization and environmental stressors, in addition to the immediate physical effects of climate change. Rapid urbanization has resulted in crowded living conditions, poor air quality, limited access to clean water, and a lack of sanitary facilities, especially in developing nations. From chronic illnesses like asthma and cardiovascular diseases to infectious diseases and malnutrition, these factors are linked to a variety of health issues.

Urban living has also been associated with an increase in mental health issues. Compared to people who live in rural areas, many studies show that city dwellers are more likely to suffer from social isolation, stress, anxiety, and depression. Combining the effects of climate change with ongoing exposure to noise, pollution, and crowded areas can lead to a stressful environment that increases health risks. As an illustration, the growing occurrence of extreme weather events not only jeopardizes people's physical safety but also increases their levels of anxiety and post-traumatic stress disorder.

Moreover, public health issues are increasingly being driven by environmental stressors like air and water pollution, unregulated urbanization that encroaches on natural habitats, as well as disruptions brought on by climate change. Vulnerable groups, such as the elderly, children, and people with underlying medical conditions, are disproportionately impacted by these stressors. These problems' convergence emphasizes how crucial it is to create urban environments that support health resilience so that communities can endure environmental stress while preserving the health and well-being of their residents.

1.2 The Concept of Climate-Resilient Communities

The idea of climate-resilient communities has gained popularity as a response to these interconnected problems. A community that anticipates, prepares for, responds to, and adapts to the negative impacts of climate change while still being able to prosper is said to be climate-resilient. Being resilient involves more than just surviving calamities; it also involves adjusting to new circumstances in ways that improve human health and environmental sustainability.

Climate-resilient design prioritizes flexibility, adaptability, and sustainability. It includes elements like energy-efficient structures, green infrastructure, and integrated health systems that can endure shocks and promote community well-being [21–24].

In addition to enhancing air quality and offering areas for social interaction and physical exercise, green roofs, permeable surfaces, and urban forests also lessen the effects of flooding and excessive heat [25]. In addition to creating healthier indoor environments, building designs that prioritize natural ventilation, daylighting, and the use of sustainable materials also lessen dependency on energy-intensive systems that worsen the effects of climate change.

To make sure that building and urban design not only addresses environmental sustainability but also safeguards and enhances community well-being, it is crucial to incorporate health considerations into climate resilience strategies. There has never been a greater need for health-resilient design as cities continue to expand and climate threats increase. It is essential to adopt this strategy to creat a prosperous, just, and sustainable future.

2 Reimagining Built Environments: Shifting Away from the Current Business-as-Usuals

It is increasingly clear that conventional building and urban development methods are inadequate to meet the evolving challenges posed by climate change, urbanization, and population growth. While the connection between the environment, public health, and social equity is becoming more widely recognized, the "business-as-usual" model dominates much of modern urban design. The complex, long-term issues we currently face are not adequately addressed by this traditional strategy, which is typified by short-term thinking, reactive planning, and an emphasis on economic growth over sustainability. As cities continue to expand and climate risks intensify, there is a pressing need to move away from outdated practices and embrace innovative, purpose-built designs that prioritize resilience, sustainability, and public health.

In order to meet short-term demands, like housing and infrastructure, traditional building and urban development methods have mostly been reactive rather than proactive, ignoring long-term sustainability and the needs of future generations. Numerous environmental, social, and health problems are aggravated by urban sprawl, unrestrained development, and inefficient building design. The design of many cities favours private vehicles over public transit, which increases air pollution, causes traffic jams, and reduces the amount of green space. Furthermore, the design of buildings that rely on energy-intensive materials and inefficient energy systems increases carbon emissions, which fuels climate change and environmental deterioration.

Low-density, car-dependent communities have proliferated due to urban sprawl in particular, placing a burden on natural resources and causing socioeconomic divisions. Marginalized communities are disproportionately impacted by these expansive developments because they frequently lack access to basic services like public transportation, healthcare, and education. There are serious health ramifications; chronic diseases like obesity, mental health issues, and respiratory disorders are all associated with poor air quality, restricted access to green spaces, and a higher risk of heat islands. The social fabric of communities is negatively impacted by urban layouts that promote isolation rather than connection, in addition to the costs to the environment and human health. The lack of accessible public areas and walkable neighbourhoods erodes social cohesiveness and makes communities less resilient to environmental or economic shocks.

The business-as-usual approach perpetuates these problems by focusing on short-term economic gain at the expense of long-term health and environmental sustainability [26, 27]. Buildings and communities are frequently planned with little consideration for their long-term viability, environmental impact, or occupants' well-being. Cost-cutting strategies usually lead to subpar insulation, infrastructure, and materials, which not only make buildings more vulnerable to climate-related risks but also make them less liveable. A scan through the global housing trend development can simply indicate such changes in terms of poor-quality living spaces, smaller spaces, low-quality construction materials, poor insulation, worsening natural lighting and ventilation, etc. Furthermore, developers frequently put profit margins ahead of building better communities, which results in unsustainable and inaccessible developments—something that we can see replicating in the Global South more than ever.

Given these difficulties, there is a strong argument for a paradigm shift in urban planning that abandons the reactive, routine approach and adopts creative, progressive methods of construction and community development. This change necessitates incorporating sustainability principles into all facets of architectural and urban design, from energy-efficient structures to natural solutions that tackle public health and climate resilience. It entails rethinking urban settings as places that promote social justice, health, and well-being while reducing their negative effects on the environment. Outdated development models must be replaced by purpose-built ones, which are made to be resilient, sustainable, and adaptive. In addition to integrating energy-efficient materials and designs that support environmental sustainability and public health, these communities should place a high priority on having access to green spaces, public transportation, and other necessities.

A fundamental rethinking of the way we construct buildings and plan cities is necessary for the future. There is a much higher demand for sustainable infrastructures and buildings. While we know the solutions, there are significant struggles to get things right! We can build urban environments that not only withstand the challenges of climate change but also support healthier, more equitable communities by putting the long-term needs of people and the planet first. Thus, it is imperative to make the shift from conventional, business-as-usual design to purpose-built, sustainable design in order to ensure a resilient future.

3 Core Principles of Health and Resilience in Design

This book, the first of our two volumes, is based on fundamental health and resilience design principles. It highlights the need for urban environments and buildings to support residents' health and well-being in addition to meeting the requirements of climate resilience. In tackling these two goals, the book highlights a number of fundamental ideas that direct the development of environments that are both sustainable and health-promoting.

Health-centric design, which puts people's physical and mental health first in both buildings and larger community settings, is a fundamental idea covered throughout the book. By emphasizing how design features—like access to natural light, ventilation, green spaces, and safe, walkable environments—can improve health outcomes, this idea goes beyond conventional approaches to architecture and urban planning. In order to create healthier living and working environments, the book explores how well-designed buildings can alleviate respiratory conditions, lower stress levels, and promote social interaction. To guarantee that the built environment promotes long-term well-being, these health-centric elements must be incorporated into building design.

The resilience principle is equally significant. Communities and structures must be built to resist and adapt to environmental stressors as the effects of climate change become more noticeable. The book examines how resilience in design includes spaces' ability to adapt to shifting environmental conditions while preserving functionality and safety, going beyond planning for extreme weather events. The book emphasizes how resilient design is essential to protecting communities from the effects of climate change, from heat-adaptive buildings to flood-resistant infrastructure. This idea also applies to social resilience, making sure that groups can remain cohesive and healthy even when things change.

The book concludes by highlighting the necessity of striking a balance between sustainability and well-being. The goal of sustainable design is frequently to minimize negative effects on the environment by using eco-friendly materials or cutting back on energy use. Nonetheless, this book promotes an integrated strategy in which community health and well-being are matched with sustainability objectives. It makes the case that settings that improve people's quality of life while simultaneously safeguarding the environment are genuinely sustainable. This principle is central to the book's vision of designing buildings and communities that are not only climate-resilient but also nurturing and supportive of human health.

4 Building a Framework for Action: Pathways to a Sustainable Future

The following chapters of the book provide a forward-looking perspective, grounded in practical strategies, at a critical juncture in the face of growing climate challenges. This aligns with the book's overarching aim to provide a thorough framework for rethinking and reimagining the built environment, where the integration of sustainability, resilience, and health serves as the basis for prosperous communities in the future. The book calls for a radical shift in how we approach architectural and community design, urging us to move away from unsustainable practices that put immediate profits ahead of long-term well-being and environmental protection.

In-depth discussions of important subjects like resilient infrastructure, health-centric design, and incorporating biophilic design are covered in the upcoming chapters. The book offers useful insights into putting theory into practice, from designing flood-resistant and heat-adaptive communities to developing energy-efficient buildings that support both physical and mental health. Each chapter equips architects, urban planners, legislators, and community leaders with the tools and resources needed to build environments that improve everyone's quality of life while also withstanding environmental stresses.

At the heart of this book's vision is the belief that adopting creative, sustainable design is not merely an option, but a necessity.. We can create new paths that support the well-being of people and the environment by departing from conventional wisdom. All parties involved—from designers to governments and communities—must be dedicated to and work together toward this shared goal. By working together, we can build a more just, healthy, and climate-resilient future [28]. Therefore, the book extends an invitation to readers to join this life-changing adventure. Now is the moment to take action. We can co-create the foundation (i.e., through better partnerships, such as those highlighted in Sustainable Development Goal 17 or SDG #17) for a sustainable future in which both people and the environment prosper by embracing the ideas presented in this book.

References

1. Allam Z, Cheshmehzangi A, Jones DS (2023) The economic landscape of cities and climate change. In Climate and social justice: the political economy of urban resilience and mercantilism. Springer Nature Singapore, Singapore, pp 25–41
2. Calanca P (2007) Climate change and drought occurrence in the Alpine region: how severe are becoming the extremes? Global Planet Change 57(1–2), 151–160
3. Cheshmehzangi A (2020) The analysis of global warming patterns from 1970s to 2010s. Atmos Clim Sci 10(3):392–404
4. Cheshmehzangi A, Dawodu A (2019) Case study reviews: people, perspective and planning. In Sustainable urban development in the age of climate change: people: the cure or curse, 69–131
5. Goodess CM (2013) How is the frequency, location and severity of extreme events likely to change up to 2060? Environ Sci Policy 27:S4–S14

6. Page EA (2007) Climate change, justice and future generations. Edward Elgar Publishing, Cheltenham
7. Sharifi A, Pathak M, Joshi C, He BJ (2021) A systematic review of the health co-benefits of urban climate change adaptation. Sustain Cities Soc 74:103190
8. Cheshmehzangi A, He BJ, Sharifi A, Matzarakis A (2023) Climate change, cities, and the importance of cooling strategies, practices, and policies. In Climate change and cooling cities. Springer Nature Singapore, Singapore, pp 2–19
9. Birkeland J (2012) Design for sustainability: a sourcebook of integrated ecological solutions. Routledge, New York
10. Birkeland J (2022) Nature positive: interrogating sustainable design frameworks for their potential to deliver eco-positive outcomes. Urban Sci 6(2):35
11. Cheshmehzangi A, Dawodu A, Sharifi A (2021) Sustainable urbanism in China. Routledge, New York
12. Gibbons LV (2020) Moving beyond sustainability: a regenerative community development framework for co-creating thriving living systems and its application. J Sustain Dev 13(2):20–52
13. Ramirez Lopez LJ, Grijalba Castro AI (2020) Sustainability and resilience in smart city planning: a review. Sustainability 13(1):181
14. Shi Y, Yuan X, Tang Y, Li Y, Wang Q, Ma Q, Zuo J, Liu H (2022) Localized regional life cycle model research for the impacts of carbon dioxide on human health and ecosystem. Sustain Prod Consumption 29:36–45
15. Un-Habitat (2008) State of the World's cities 2008/9: Harmonious Cities. Earthscan, Oxford
16. Zuo J, Shi M, Jia W, Yan Y, Song X, Shen R, He J (2024) Reducing heat exposure from personal cooling strategies to green city construction in China's tropical city. Sci Total Environ 927:171955
17. Ziafati Bafarasat A, Cheshmehzangi A, Ankowska A (2023) A set of 99 healthy city indicators for application in urban planning and design. Sustain Dev 31(3):1978–1989
18. Zuo J, Pullen S, Palmer J, Bennetts H, Chileshe N, Ma T (2015) Impacts of heat waves and corresponding measures: a review. J Clean Prod 92:1–12
19. Hu Y, Lin Z, Jiao S, Zhang R (2023) High-density communities and infectious disease vulnerability: a built environment perspective for sustainable health development. Buildings 14(1):103
20. Pantelic J, Aristizabal S, Liu Q, Senerat A, Son YJ, Byun K, Li L, Mullan A, Zhang R, Johnson BD, Clements N (2023) The impact of automated control of indoor air pollutants on cardiopulmonary health, environmental comfort, sleep quality in a simulated apartment: a crossover experiment protocol. Frontiers in Built Environment 9:1117992
21. Erten D (2024) Sustainable green buildings and infrastructure for climate resilience. Handbook of climate change impacts on river basin management. CRC Press, Boca Raton, FL, pp 269–284
22. Infield EMH, Abunnasr Y, Ryan RL (eds) (2018) Planning for climate change: a reader in green infrastructure and sustainable design for resilient cities. Routledge, New York
23. Kumareswaran K, Jayasinghe GY (2023) Green infrastructure and urban climate resilience. Springer, Cham
24. Sharifi A (2016) A critical review of selected tools for assessing community resilience. Ecol Ind 69:629–647
25. Gao J, Kovats S, Vardoulakis S, Wilkinson P, Woodward A, Li J, Gu S, Liu X, Wu H, Wang J, Song X, Zhai Y, Zhao J, Liu Q (2018) Public health co-benefits of greenhouse gas emissions reduction: a systematic review. Sci Total Environ 627:388–402
26. Beckmann M, Schaltegger S (2021) Sustainability in business: integrated management of value creation and disvalue mitigation. In Oxford research encyclopedia of business and management
27. York JG (2009) Pragmatic sustainability: Translating environmental ethics into competitive advantage. J Bus Ethics 85:97–109
28. Ziafati Bafarasat A, Sharifi A (2024) How to Achieve a Healthy City: a Scoping Review with Ten City Examples. J Urban Health 101(1):120–140

Architectural Design Practices for Healthy Buildings

Building Well at Delos Office: Work/Life Elevated on Manhattan's West Side High Line

Regina Vaicekonyte, Sara Aristizabal, and Anna Obraztsova

Abstract This case study explores the design and construction of Delos Global Headquarters at 860 Washington Street in New York City, a cutting-edge example of integrating health, wellness, and sustainability in the built environment. The building, situated above the High Line, epitomises the intersection of modern architecture and progressive building certification standards, including WELL New & Existing Interiors, LEED v4, and Living Building Challenge Petal Certification. Delos HQ serves as a model for "Building-level Practices" by applying the 10 concepts of the WELL Building Standard™ (WELL)—Air, Water, Nourishment, Light, Movement, Thermal Comfort, Sound, Materials, Mind, and Community—each playing a part in a holistic strategy for occupant health and well-being. Methodologically, the case study examines the incorporation of advanced HVAC systems, underfloor air distribution, and MERV 13 filters to optimise air quality and thermal comfort among employees throughout the office space. It also evaluates the use of biophilic design elements, circadian lighting systems, and sound-absorbing materials to enhance both physical and mental comfort. The office layout promotes movement through features like open stairways and interactive digital screens, while ergonomic workstations and a variety of focus rooms cater to different work styles. At a community level, Delos HQ strengthens its connection to the neighbourhood through its industrial design inspired by the High Line and a terrace that supports local biodiversity. In addition, the building demonstrates its commitment to sustainability through its recycling and composting practices, the use of energy-efficient appliances, and the incorporation of reclaimed wood. Collectively, these building and community-level interventions foster a holistic environment that promotes occupant health, productivity and sustainability, and through rigorous adherence to WELL requirements, exemplify how innovative design and operational practices can set new benchmarks in building excellence.

Keywords WELL Building Standard · Indoor environmental quality · Indoor air quality · Water quality · Circadian lighting and daylight · Biophilic design

---

R. Vaicekonyte (✉) · S. Aristizabal · A. Obraztsova
Delos Living LLC, New York, NY, USA
e-mail: regina.vaicekonyte@gmail.com

A. Cheshmehzangi et al. (eds.), *Designing Healthy Buildings and Communities*, Urban Sustainability, https://doi.org/10.1007/978-981-96-3057-8_2

1 Introduction

Situated in the dynamic Meatpacking District, above the High Line, the Delos Global Headquarters symbolises the modern and progressive architecture of the evolving area and exemplifies leading-edge building certification standards. This area, where historic urban elements blend with tech innovations and upscale fashion boutiques, provides a fitting backdrop for Delos's mission to "Innovate Well". Built with a glass façade, the building offers views of Manhattan's downtown skyline, the Hudson river and the hustling and bustling foot traffic of the High Line—all of which are made easy to see through the glass curtain walls.

Delos has been prioritising human health and well-being in the built environment since the launch of the WELL Building Standard in 2014 [1]. Serving as a valuable case study in sustainable and healthy design, the Delos HQ aimed to meet ambitious health and wellness objectives while also pushing the envelope in sustainability and green building design, leading to the pursuit of three building certifications: WELL New & Existing Interiors, LEED v4 [2] and Living Building Challenge Petal Certification [3].

The interior fit out construction of the 19,318 square foot new office space was completed in late 2017. It spans the 4th and 5th floors, and includes an open plan office space, meeting rooms of varying sizes, several private offices and a few huddle/focus rooms. The fourth floor also has a 1,520-square-foot outdoor terrace that features plants native to New York state and offers tables and seating for both individual work and small meetings as well as lunch and quick breaks.

In this case study, we provide an overview of the WELL Building Standard and how the construction of Delos' 860 Washington Street HQ addressed it, in addition to LEED v4 [2] and Living Building Challenge Petal Certification [3] Requirements (Fig. 1).

2 The WELL Building Standard

The WELL Building Standard, or WELL, is grounded in years of interdisciplinary research and serves as a framework for creating spaces that promote human health and well-being. WELL v2 introduces a series of strategies, informed by the latest scientific evidence, aimed at improving health through design changes, operational practices and policies, while fostering a culture centred around health and wellness. Building on the foundation laid by the first version of the WELL Building Standard (WELL v1) [1], WELL v2 [4] incorporates insights from a global community of WELL users, practitioners, public health experts and building scientists.

WELL is structured around 10 core concepts, each containing features specifically designed to address various health-related objectives.

Fig. 1 Outdoor terrace at Delos office in the Meatpacking District of New York City

1. Air: The WELL Air concept aims to maintain high indoor air quality throughout a building's life by employing various strategies, including eliminating or reducing pollution sources, optimizing building design and operations (e.g., enhancing ventilation, filtration, air quality monitoring, using low-emission materials and installing air purification systems), as well as promoting behavioural changes (e.g., smoking bans).
2. Water: The WELL Water concept focuses on the quality, distribution and management of water within a building. It includes features ensuring access to clean drinking water, setting contaminant limits, and managing water use to protect both the building's structure and the surrounding environment.
3. Nourishment: The WELL Nourishment concept ensures the availability of fresh fruits and vegetables and promotes the nutritional transparency of food served or sold within the building. It advocates for mindful eating, nutrition education and creating environments where making healthy food choices is the easiest option.
4. Light: The WELL Light concept encourages exposure to beneficial lighting and designs spaces that enhance visual, mental and biological health. Requirements include providing lighting at workstations, simulating natural daylight and allowing occupants to adjust lighting in their immediate environment.

5. Movement: The WELL Movement concept encourages physical activity through thoughtful design, policies and programs that integrate movement opportunities into buildings and communities. It includes provisions such as adjustable workstations, active furniture, designated spaces for exercise and facilities supporting active commuting.
6. Thermal comfort: The WELL Thermal Comfort concept promotes productivity and comfort by improving HVAC system design, allowing for temperature adjustments and meeting individual thermal preferences. Key requirements include thermal zoning and enabling occupants to control temperature in their immediate environment.
7. Sound: The WELL Sound concept enhances occupant health and well-being by identifying and addressing acoustical factors that influence comfort. It includes the creation of acoustic plans to reduce disruptive noise, enhance sound privacy, use sound-absorbing materials and minimize impact noise.
8. Materials: The WELL Materials concept seeks to reduce exposure to harmful chemicals that may affect health during the construction, renovation and operation of buildings. Requirements include limiting hazardous materials, providing transparency about the materials used and implementing cleaning protocols that avoid toxic chemicals.
9. Mind: The WELL Mind concept enhances mental well-being by implementing supportive policies, programs and design elements that address cognitive and emotional health. Features include mental health education, access to nature, stress management programs, restorative spaces and substance use education and services.
10. Community: The WELL Community concept fosters a supportive environment by ensuring access to critical healthcare and creating a culture that meets diverse needs. Requirements include policies for new parents, emergency resources (e.g., first aid kits, AEDs) and regular occupant surveys to assess satisfaction with the space.

WELL v2 is based on the following principles:

- Equitable: Aims to benefit all populations, especially disadvantaged or vulnerable groups.
- Global: Offers strategies that are feasible and relevant worldwide.
- Evidence-based: Relies on rigorous research across multiple disciplines, validated by a global network of experts.
- Technically sound: Defines best practices and outlines approaches through performance verification and third-party assessments.
- Customer-focused: Supports WELL users via coaching, dynamic resources and an intuitive platform for navigating the certification process.
- Resilient: Adapts to advances in research, science, technology and societal needs, continuously integrating new findings to stay current.

The WELL Building Standard evaluates buildings based on the 10 concepts described above. Each concept contributes to a building's overall score, with specific features within each concept offering points based on their implementation and impact. The total points accumulated across these concepts determine the building's certification level, with the four levels being WELL Bronze, WELL Silver, WELL Gold and WELL Platinum. These levels reflect the degree to which a building promotes the health and well-being of its occupants, with WELL Platinum representing the highest standard of excellence. The rating system encourages comprehensive strategies that address both the physical and psychological aspects of a healthy indoor environment.

As of March 2025, the WELL Building Standard has been implemented across more than 5.79 billion square feet of space in 137 countries. The standard has been adopted by over 150 companies from the Fortune 500 and Global 500 lists, as well as many others—enhancing the health and well-being of an estimated 25 million individuals in over 74,000 commercial and residential buildings [5].

3 Case Study: Delos Office Design and the 10 Concepts of WELL

As the pioneer of Wellness Real Estate® and the founder of the WELL Building Standard, Delos prioritizes occupant health and well-being in the built environment. This commitment has been the most evident in Delos's headquarters in New York City. The following sections describe how the 10 concepts of WELL were incorporated at every phase, from design and construction to occupancy.

3.1 Air

The Delos HQ is designed with an underfloor air distribution (UFAD) system that employs a natural temperature gradient to deliver clean air from the floor and expel polluted air through the ceiling [6] (Figs. 2 and 3). This design prioritises thermal comfort and strategic diffuser placement. Walk-off grates at the building's entrance and in front of elevators reduce the ingress of outdoor particulate matter, enhancing indoor air quality. Additionally, floor grates on the terrace contribute to localised air cleanliness.

The headquarters are also equipped with remote working and conferencing technologies, as well as a large digital display showcasing Delos's mission and impact, along with real-time indoor environmental quality (IEQ) monitoring. Environmental sensors monitor air quality continuously and display visual alerts when outdoor air quality thresholds are exceeded.

Building materials, furniture and even cleaning products in the space were selected for their low volatile organic compound (VOC) emissions. The air was tested for contaminants such as VOCs, formaldehyde and ozone during and after construction. Additionally, sensors installed throughout the space continuously gather data on

Fig. 2 Diffusers of underfloor air distribution systems

Fig. 3 Underfloor air distribution system diffusers in the Delos office.

building performance and track indoor air quality to maintain ongoing compliance to IAQ standards, facilitating prompt responses to deviations and enhancing occupant awareness [7].

The project meets and exceeds ASHRAE 62.1 ventilation standards [8] by 30%, ensuring abundant fresh air circulation. For outdoor air, the mechanical system uses

MERV 13 filters and can accommodate carbon filters for VOC removal. In densely occupied office areas, demand-controlled ventilation maintains CO_2 levels below 800 ppm. During construction, SMACNA guidelines were followed to minimise particulate matter, and an air flush was conducted post-construction to expel polluted air.

Fifteen active green walls throughout both floors (NAAVA, Helsingfors, Finland)—present in nearly every room and space—also help reduce indoor VOCs, in addition to providing a sense of connectedness with nature. The green walls are low-maintenance, with automatic watering systems and built-in fans that pull air across plant roots embedded in a carbon medium, breaking down VOCs in the process.

3.2 Water

At Delos HQ, water is filtered through a multi-stage system that includes carbon, sediment and microbial cyst removal to ensure clean and palatable drinking water. All sinks are designed to provide ample space for comfortable handwashing, and the soap meets EcoLogo/GreenSeal standards, is fragrance-free and is supplied in non-refillable sealed cartridges.

Water dispensers are strategically placed on both floors to provide convenient access to clean drinking water, encouraging regular hydration among employees. These stations undergo regular maintenance and cleaning to ensure consistently good water quality.

3.3 Nourishment

Delos offers a variety of healthy food options, including organic fruits and vegetables, as well as snacks and beverages containing less than 25 g of sugar per serving, all labelled with detailed nutritional information [9]. Cooking tools are constructed from non-contaminating materials, and portion sizes are regulated using appropriately sized plates, bowls and cups. Additionally, snacks and beverages are sourced from local farms whenever feasible.

The office features an outdoor terrace with distinct zones designed for flexible use, providing ample space for mindful eating. The café and terrace areas provide sufficient seating to accommodate at least 25% of employees, allowing them to enjoy lunch away from their workstations.

Fig. 4 Daylight access through floor to ceiling windows and circadian lighting system

3.4 Light

Daylight levels are optimised using interior sheer shading [10]. The circadian lighting system is programmed to adjust brightness and correlated colour temperature in sync with the solar day, supporting employees' circadian rhythms (Fig. 4). Additionally, task lighting at individual workstations allows for precise adjustment of light intensity and colour temperature [11].

The design of the office ensures that 75% of the floor area is within 25 feet of fenestrations and 95% of workstations are situated within 41 feet, ensuring uniform distribution of natural light and unobstructed views for all occupants.

3.5 Movement

The open stairway is designed to encourage movement between floors, featuring extensive views, natural daylight, artwork and biophilic elements (Fig. 5). Interactive digital screens provide real-time feedback and incentives for stair use in the form of digital art, thereby promoting physical activity. Height-adjustable workstations and ergonomic chairs support optimal posture while working, and bike chairs and treadmill desks provide employees the opportunity to be physically active while working.

The free-address workspace design allows employees to select their preferred work areas based on environmental parameters such as temperature and acoustics. The office layout includes various focus rooms and flexible meeting spaces to accommodate different work styles and requirements.

Fig. 5 An interactive digital screen next to the stairs encourages occupants to increase physical activity

3.6 Thermal Comfort

The HVAC system at Delos HQ is engineered to create a thermal gradient across the workspace, enabling occupants to select areas with their preferred thermal conditions. This design promotes individual comfort by allowing employees to move to zones where the temperature aligns with their personal preferences.

3.7 Sound

The acoustical ceiling features a Noise Reduction Coefficient (NRC) of 0.9 which effectively absorbs sound. This is further complemented by additional sound-absorbing materials, including perforated wood panelling and specialised acoustical panels, which enhance acoustical comfort in both offices and conference rooms. These elements work together to minimise noise and improve overall sound quality within the workspace (Fig. 6).

Fig. 6 Sound absorption panel in a closed meeting room

3.8 Materials

All materials throughout the office space were carefully selected for their health- and sustainability-related properties, with an emphasis on low VOC content and compliance with the Living Building Challenge (LBC) Red List [12]. Cleaning protocols ensure the use of EcoLogo or GreenSeal certified products, HEPA-filter vacuums and microfiber rags.

Specific product successes include PVC-free and isocyanate-free Interface carpet, Red List compliant Superior Essex cabling, no added urea formaldehyde wood, and compliant alternative edge-banding from Teknion, PVC-free Koncept task lights and Red List-free tape for signage (Table 1).

3.9 Mind

Access to plants and natural patterns, both indoors and outdoors, supports mental health [13]. The open office design incorporates biophilic elements, including views of the Hudson River, High Line greenery, plant walls throughout both floors of the office and nature-inspired patterns incorporated into the office design. The terrace offers direct connection to nature, while the wellness room provides a space for recharging, equipped with warm lighting, a reclining chair, a lock, a sink and a refrigerator, essential for nursing mothers. Everyone in the space is encouraged to engage with nature both indoors and on the outdoor terrace, creating a welcoming and productive environment (Fig. 7).

Table 1 The 2024 LBC Red List highlights a list of chemicals considered the "worst in class" substances commonly found in the building industry, which pose significant risks to human health and the environment

2024 red list chemical class changes (since April 2023)			
Chemical class	Red list (RL)	Priority list (PL)	Watch list
Alkylphenols and related compounds	374 CASRNs added from PL	69 CASRNs added from WL 24 CASRNs d fleetly added 1CASRN corrected" 2 duplicate CASRNs removed	All CASRNs advanced to PL
Antimicrobials (marketed with a health claim)	No changes	No changes	No changes
Asbestos compounds	12 CASRNs added from PL	All CASRNs advanced to RL	No changes
Bisphenol A (BPA) and structural analogues	25 CASRNs added Trom PL	25 CASRNS advanced to RL 1 CASRN directly added 2 CASRNs corrected	No changes
California-banned solvents	No changes	No changes	No changes
Chlorinated Polymers	111 CASRNs added from PL	111 CASRNs advanced to RL 2 duplicate CASRNs removed	12 duplicate CASRNs removed
Chlorobenzenes	No changes	1 CASRN directly added	No changes
Chlorofluorocarbons (CFCs) and hydrochlorofluorocarbons (HCFCS)	No changes	2 CASRNS directly added	No changes
Formaldehyde (added)	1 CASRN removed	389 CASRNs added 1 T: m WL 734 CASRNs directly added	All CASRNs advanced to FL 1 CASRN added From RL 2 duplicate CASRNs removed
Monomeric, polymeric, and organophosphate halogenated flame retardants (HFR5)	189 CASRNs added from PL	189 CASRNs advanced to RL 63 CASRNs added From WL 1119 CASRNs directly added 4 duplicate CASRNs removed	All CASRNs advanced to FL 26 duplicate CASRNs removed

(continued)

Table 1 (continued)

2024 red list chemical class changes (since April 2023)			
Chemical class	Red list (RL)	Priority list (PL)	Watch list
Organotin compounds	No changes	202 CASRNs added from WL	All CASRNs advanced to FL
Per- and Polyfluorinated Alkyl Substances (PFA5)	32 CASRNs corrected	31 CASRNs directly added	13 duplicate CASRNs removed
Phthalates (orthophthalates)	13 CASRNS added from PL	13 CASRNS advanced to RL. 557 CASRNs directly added 1 duplicate CASRN removed	No changes
Polychlorinated biphenyls (PCBs)	No changes	No changes	No changes
Polycyclic aromatic hydrocarbons (PAHs)	70 CASRNs added from PL	70 CASRNs advanced to RL	No changes
Short-chain and medium-chain chlorinated paraffins	1 CASRN added from FL	1 CASRN advanced to the RL	No changes
Toxic heavy metals	122 CASRNs added from PL	122 CASRNs advanced to RL 2476 CASRNs added from WL 904 CASRNs directly added	All CASRNs advanced to PL 28 duplicate CASRNs removed
Volatile organic compounds (VDCs) In wet-applled products	No changes	No changes	No changes
Wood treatments containing creosote or pentachlorophenol	No changes	103 CASRNs directly added	1 duplicate CASRN remaned
C1-C4 halogenated hydrocarbons	No changes	No changes	344 CASRNs added
Asphalt	No changes	No changes	2 CASRN added
Stoddard solvent	No changes	No changes	1 CASRN added
Isocyanates	No changes	No changes	24 CASRNs added

Volatile organic compounds (VOCs) In on-site wet-applled products are not banned, but must have VCC levels below the South Coast Air Quality Management District (SCAQMD Rule 1168 for Adhesives and Sealants or the CARB 2007 Suggested control Measure (SCM) for Architectural Coatings, as applicable

Fig. 7 Biophilic design elements aimed at supporting mental health and well-being

3.10 Community

The terrace, featuring native plants such as milkweed, supports local biodiversity and attracts monarch butterflies. The office's industrial design, inspired by the High Line, offers quality views of Manhattan and strengthens the connection to the local community.

3.11 Reducing Environmental Impact

Delos HQ prioritises sustainability through daily recycling and composting, rigorous construction waste management and energy-efficient appliances. Water usage is minimised with low-flow faucets and fixtures, and reclaimed wood is used throughout the space.

4 Concluding Remarks

This case study represents the state of Delos Global Headquarters at the time of initial occupancy in late 2017 and through 2020. During and after the Covid-19 pandemic, Delos has strategically repositioned the use of this office to best support the hybrid in-person and remote working mode of the business.

The office stands as a pioneering example of integrating health and well-being into an indoor space. By following the WELL Building Standard, Delos has created a workspace that not only prioritises sustainability but also enhances the overall health and wellness of its occupants.

Throughout the design and construction of their new office, Delos meticulously implemented strategies to ensure optimal air quality and water filtration. Advanced HVAC systems promote thermal comfort and reduce indoor air pollutants, while dynamic lighting and sound-absorbing materials support a comfortable and productive work environment.

Delos HQ serves as a model for future developments, demonstrating that innovative design and operational practices can foster environments that enhance the health, wellness and productivity of occupants, while also adhering to strict sustainability standards.

References

1. International WELL Building Institute (2014) WELL Building Standard v1. International WELL Building Institute, New York, NY. Retrieved from https://www.wellcertified.com/the-standard/v1
2. Green Building Council (2019) LEED v4 for building design and construction. U.S. Green Building Council. Retrieved from https://www.usgbc.org/resources/leed-v4-building-design-and-construction-current-version
3. International Living Future Institute (2022) Living Building Challenge. Retrieved July 29, 2024, from https://living-future.org/lbc
4. International WELL Building Institute (2023) WELL Building Standard v2. International WELL Building Institute, New York, NY. Retrieved from https://v2.wellcertified.com/en/wellv2/overview
5. International WELL Building Institute (2025) WELL Certified. Retrieved March 1, 2025, from https://www.wellcertified.com/
6. Zhang K, Zhang X, Li S, Jin X (2014) Review of underfloor air distribution technology. Energy Build 85:180–186. https://doi.org/10.1016/j.enbuild.2014.09.011
7. International WELL Building Institute (2020) Ventilation effectiveness. Retrieved July 29, 2024, from https://standard.wellcertified.com/v3/air/ventilation-effectiveness
8. American Society of Heating, Refrigerating and Air-Conditioning Engineers (2019) ANSI/ASHRAE Standard 62.1–2019: Ventilation for Acceptable Indoor Air Quality. ASHRAE. https://www.ashrae.org/technical-resources/bookstore/standards-62-1-62-2
9. International WELL Building Institute (2020) Processed foods. Retrieved July 29, 2024, from https://standard.wellcertified.com/nourishment/processed-foods
10. Jamrozik A, Clements N, Hasan SS, Zhao J, Zhang R, Campanella C, Loftness V, Porter P, Ly S, Wang S, Bauer B (2019) Access to daylight and view in an office improves cognitive performance and satisfaction and reduces eyestrain: a controlled crossover study. Build Environ 165:106379. https://doi.org/10.1016/j.buildenv.2019.106379 https://www.sciencedirect.com/science/article/pii/S036013231930589X
11. International WELL Building Institute (2020) Circadian lighting design. Retrieved July 29, 2024, from https://standard.wellcertified.com/light/circadian-lighting-design
12. Living Building Challenge Red List. Living-Future.org. https://living-future.org/red-list/. Accessed July 24, 2024.U.S.
13. Aristizabal S, Byun K, Porter P, Clements N, Campanella C, Li L, Mullan A, Ly S, Senerat A, Nenadic IZ, Browning WD, Loftness V, Bauer B (2021) Biophilic office design: Exploring the impact of a multisensory approach on human well-being. J Environ Psychol 77: 101682. https://doi.org/10.1016/j.jenvp.2021.101682 https://www.sciencedirect.com/science/article/pii/S0272494421001353

Characterization of Healthy Buildings in Yazd Traditional Houses

Saeedeh Moayedfar and Seyedeh Saeedeh Hoseinyzadeh

Abstract In the face of escalating industrialization and climate change, achieving sustainability and promoting its development in contemporary urban environments is paramount. Healthy Buildings, characterized by their prioritization of occupant health, well-being, and minimal environmental impact, are fundamental to constructing sustainable cities. By focusing on strategies that reduce energy and water consumption and align with sustainability principles, Healthy Buildings contributes to thriving and resilient urban environments. Arid cities, particularly vulnerable to the challenges posed by climate change, can benefit significantly from traditional building practices. Traditional Houses in Yazd, renowned for their resilience and health-promoting features, offer valuable examples of climate-resilient design. This research employs a mixed-methods approach, combining qualitative and quantitative methodologies, to investigate the alignment of traditional Yazd houses with contemporary Healthy Building criteria. Key research indicators were identified through content analysis, and their scores and weights were evaluated using the Analytic Hierarchy Process (AHP). By examining these centuries-old resilient structures, this study aims to promote traditional Yazd houses as models for new construction. Findings indicate that a significant proportion of Healthy Building criteria are met by most traditional Yazd houses, positively impacting occupant health and well-being. This research contributes to a deeper understanding of Healthy Buildings in historic urban areas, fostering greater appreciation for these spaces and a deeper understanding of climate resilience over time.

Keywords Climate resilience · Arid climate · Healthy building · Traditional houses · Yazd City

---

S. Moayedfar (✉)
Department of Geography, Faculty of Humanities, Meybod University, Meybod, Iran
e-mail: moayedfar@meybod.ac.ir

S. S. Hoseinyzadeh
Department of Art and Architecture, Payame Noor University, Tehran, Iran

A. Cheshmehzangi et al. (eds.), *Designing Healthy Buildings and Communities*, Urban Sustainability, https://doi.org/10.1007/978-981-96-3057-8_3

1 Introduction

Global societies face enormous challenges due to the escalating effects of climate change, typified by rising temperatures, harsh weather, and rising sea levels [1]. Cities are particularly vulnerable to these impacts as primary hubs of human activity and economic growth. Consequently, a growing imperative exists to design and manage urban environments that can withstand and adapt to climate change.

The concept of climate-resilient cities has emerged as a framework for addressing these challenges. Such cities prioritize integrating climate considerations into urban planning, design, and management. By fostering resilience, cities can better protect their populations, infrastructure, and economies from the adverse effects of climate change [2–5]. Creating healthy and sustainable built environments is central to building climate-resilient cities. Healthy buildings protect occupants from environmental hazards and enhance their physical, mental, and social well-being [6, 7]. By incorporating principles of sustainable design and prioritizing occupant health, Healthy Buildings can significantly contribute to the overall resilience of urban areas [8–11].

While often characterized as inherently resilient and healthy due to their historical endurance [12], traditional housing warrants a more nuanced perspective. These dwellings frequently incorporate passive design elements that unintentionally contribute to disaster risk reduction [13]. Features such as natural ventilation, solar control, and rainwater harvesting enhance prevention, mitigation, response, and recovery efforts. This underscores the value of integrating traditional knowledge with modern design principles to build resilience. Revitalizing traditional knowledge of urban planning and management is challenging but essential. We can significantly improve urban safety and resilience by harnessing its potential to reduce disaster risks. A paradigm shift is necessary: from viewing heritage as a passive victim to recognizing its active role in reducing vulnerabilities and fostering sustainable recovery [14–18].

Heritage buildings contribute to healthy built environments by preserving cultural heritage and promoting health standards. Their emphasis on local materials, natural ventilation, and sustainable conservation aligns with contemporary Healthy Building principles. By examining the historical development of traditional buildings through a public health lens, we can identify key indicators of "Healthy Buildings" and elevate them as models for modern construction practices. This study aims to demonstrate traditional buildings' potential to inform sustainable and resilient urban development practices. To achieve this goal, the research will address the following questions:

- How well do the principles of vernacular architecture in challenging climates meet the standards of Healthy Buildings?
- Can the vernacular architecture of traditional houses in Yazd, a climate-resilient city, be adopted as a viable model for modern architecture?

2 Research on Healthy Buildings

2.1 *Definition of Healthy Building*

Since the term "Healthy Building" was first introduced, various definitions have emerged, reflecting the evolving understanding of what constitutes a healthy environment. A Healthy Building supports the health and well-being of its users and inhabitants throughout the building's lifetime, from design and construction to destruction and recycling, according to the International Indoor Healthy Building Association (IIHA) (2006). This is accomplished through optimizing elements, including thermal comfort, access to natural light and vistas, safety and security, water and indoor air quality, ergonomics, and residents' psychological and social well-being. Similarly, the Building Research Establishment (BRE) (2008) describes a Healthy Building as one that is designed, constructed, used, and maintained in a manner that enhances the health, well-being, and productivity of its occupants while minimizing environmental impact. In line with this idea, the Singapore Building and Construction Authority (BCA) (2013) defines a healthy building as promoting the environment's and its occupants' health and well-being via careful planning, execution, and upkeep.

In addition to providing the required amenities and services, the Architectural Society of China [19] contributes to the conversation by suggesting that Healthy Buildings promote people's physical and emotional well-being while creating a healthier environment. This ultimately leads to better health outcomes. Ye et al. [5] take a more nuanced approach, defining healthy buildings as places that incorporate both physical and subjective aspects. This includes not only measurable factors such as temperature, ventilation efficiency, noise levels, illuminance, and air quality but also subjective psychological elements such as floor plan, space layout and color, privacy provisions, landscaping, and material selection, as well as job satisfaction and interpersonal relationships.

Carmichael et al. [20] reinforce this comprehensive understanding of a "Healthy Building." They emphasize the necessity of considering not only the physical attributes of the indoor environment but also the broader social, health, and environmental impacts on occupants and the surrounding community. This perspective highlights the importance of informing building policies and regulations with the latest public health evidence and best practices.

Similarly, [21] emphasize that this comprehensive definition of a "Healthy Building," particularly within the Chinese context, necessitates addressing not only the physical characteristics of indoor environments but also the broader social, psychological, and environmental criteria that influence the overall health and well-being of building occupants. Felicioni et al. highlighted the importance of integrating sustainability and resilience into building design to reduce vulnerability, emphasizing the need to consider both aspects from the early stages of the design process [22]. A Healthy Building encompasses more than just structural integrity. There are two primary ways to define a Healthy Building:

- **Environmentally-focused**: This definition highlights the importance of sustainable building practices and minimizing environmental impact. It includes using recycled materials, ensuring energy and water efficiency, and avoiding harmful chemicals in construction.
- **Occupant-focused**: This definition prioritizes the health and well-being of the building's occupants. It emphasizes the need for good indoor air quality, adequate ventilation, comfortable lighting and temperature, and features that promote physical activity and mental well-being.

Ideally, a truly healthy building integrates these perspectives, benefiting both the environment and the people who live, work, or spend time within its walls.

2.2 *Differences Between Healthy Buildings and Other Types of Buildings*

Various building concepts, such as green, low-carbon, sustainable, and eco-building, have been the subject of extensive research both domestically and internationally, each with its specific focus. However, none of these concepts can be equated with Healthy Buildings in terms of promoting building health. Green building, an idea rooted in arcology, is a foundational framework linked to the health of structures [23, 24]. The primary objective of this approach is to reduce energy consumption and protect the environment [25]. In this context, [26] describes green buildings as those that exist in harmony with nature, contributing to reductions in environmental pollution while maximizing the efficiency of resource use—be it energy, land, water, or materials—throughout their entire life cycle.

Green building is primarily characterized by its emphasis on curtailing energy consumption and emissions [27–29]. Moreover, such structures offer enhanced occupant well-being and a substantial return on investment for both developers and the surrounding community [30]. Following the rise of the low-carbon economy, low-carbon buildings have gained prominence [31]. These structures aim to minimize fossil fuel use and enhance efficiency throughout their life cycles [32]. By transforming the design, construction, management, and operation processes, they significantly cut carbon dioxide emissions [33]. Fuel conservation is widely recognized as essential for advancing low-carbon buildings. Additionally, the objective is to create more comfortable indoor spaces while reducing CO_2 emissions and waste through using low-carbon materials and innovative technologies [34]. Ultimately, the primary emphasis of low-carbon buildings is on lowering carbon emissions [35].

Sustainable buildings are conceived with a holistic approach, integrating environmental, social, and economic considerations aligned with the principles of sustainable development [36]. This approach similarly integrates resources, environmental concerns, and ecological awareness [37], aiming for long-term energy conservation and pollution reduction. Additionally, the movement towards sustainable building practices is increasingly influencing methods in planning, design, specification, and procurement [38]. To a certain extent, these sustainable structures share similarities with green buildings, particularly in their design, function, and overarching development goals [39].

Eco-buildings serve as a vital link between ecology and architecture. They exemplify the integration of resources, environmental stewardship, and ecological principles [40]. According to [41], eco-buildings are tasked with fostering a balanced living environment while ensuring a responsible approach to natural resource utilization. The primary focus of eco-buildings is to enhance their ecological advantages while minimizing environmental harm, guided by the local natural context [42]. Furthermore, much of the research on eco-buildings has centered on green technologies, ecological benefits, and energy efficiency [10].

Healthy Buildings are designed holistically, focusing on creating environments that promote occupant health and well-being. In contrast, previous buildings may prioritize efficiency or aesthetics without addressing the critical health aspects of the indoor environment. The focus on air quality, thermal comfort, lighting, acoustics, biophilic design, and overall occupant well-being sets Healthy Buildings apart as a necessary evolution in architecture and urban planning.

3 Research Methods

This study employs a mixed-methods approach, integrating both qualitative and quantitative methodologies to provide a comprehensive analysis of the research problem.

3.1 Qualitative Method: Content Analysis

The qualitative aspect of this research utilizes content analysis as the primary method. Content analysis is a systematic technique for analyzing textual, visual, or audio content to identify patterns, themes, and meanings. This method allows for in-depth data exploration, providing insights into the underlying messages and contexts.

The steps involved in the content analysis are as follows [43]:

1. **Data Collection:** Gather relevant materials such as articles, interviews, and multimedia content related to the research topic.
2. **Coding:** Develop a coding scheme to categorize the data into themes and subthemes.
3. **Analysis:** Analyze the coded data to identify recurring patterns and significant insights.
4. **Interpretation:** Interpret the findings in the context of the research questions, drawing connections between the themes and the broader implications of the study.

3.1.1 Key Dimensions of Healthy and Resilient Buildings by Content Analysis

The concept of a Healthy Building emerged in 1980, initially presented by a Japanese architect in the journal "New Architecture" [21]. This pioneering work laid the foundation for a burgeoning discipline devoted to designing and constructing buildings that prioritize the health and well-being of their occupants [44]. The concept gained popularity as public awareness of the built environment's impact on human health and well-being increased.

The WELL standard, created and implemented by the International WELL Building Institute (IWBI) in the United States, assesses seven essential factors required to promote human health and well-being in the built environment: air, water, food, light, fitness, comfort, and mind (IWBI, 45). In 2017, the Architectural Society of China (ASC) introduced the "Healthy Building Evaluation Standards," which address similar themes such as air, water, comfort, fitness, humanities, and services.

To determine the factors influencing Healthy Buildings, a comprehensive search was conducted on databases like Scopus and Web of Science using keywords such as "Healthy Building," "Healthy Housing," "Healthy Home," "Green Building," and others related to sustainable and healthy environments. After several filtering stages, 100 relevant articles published between 1995 and 2023 were analyzed. To supplement this data, seven experts in construction and architecture were interviewed to identify factors affecting traditional buildings in Yazd. As Iran lacks a specific standard for Healthy Buildings, three dimensions (design, construction, performance) with 18 factors contributing to the health of traditional Yazd structures were identified through a combination of literature review and expert input, as detailed in Table 1.

3.2 Quantitative Method: Analytic Hierarchy Process (AHP)

The Analytic Hierarchy Process (AHP) is employed for the quantitative component. AHP is a structured technique for organizing and analyzing complex decisions based on mathematics and psychology [47]. Among various multi-criteria decision-making (MCDM) methodologies, the analytical hierarchy process (AHP) is favored for its accuracy and effectiveness in delivering reliable results. AHP, first introduced by Saaty in 1980, is a measurement theory that employs pairwise comparisons derived from expert opinions [48]. This method assists decision-makers and managers by breaking down complex problems, allowing for comparative judgments and the synthesis of priorities [49].

Dyer and Forman highlighted several advantages of AHP (1992): (I) it effectively addresses issues with tangible, intangible, individual, and shared values; (II) it helps decision-makers focus on objectives rather than alternatives; and (III) it facilitates a systematic consideration of each factor through structural hierarchies [50]. Utilizing AHP in decision-making not only enhances the rationality and comprehensiveness of a firm's approach but also improves outcomes.

Table 1 Architectural factors influencing health in historical Yazd buildings

Dimension	Factor
Design	Site plan
	Structural design
	Architectural design
	Cost plan
	Building envelope
	Ecological protection
	Renewable energy
	Material recycling
Construction	Construction Safety
	Air pollution
	Noise pollution
	Light pollution
	Waste minimization
Performance	Living environment
	Natural environment
	Neighborhood
	Thermal quality
	Lighting quality

Source The review of the research literature, expert opinions, and emphasis on specific articles such as [24, 46]

The steps involved in AHP include:

1. **Problem Definition:** Clearly define the decision problem and the goal of the analysis.
2. **Hierarchy Structuring:** Break down the problem into a hierarchy of sub-problems, which can be analyzed independently.
3. **Pairwise Comparison:** Conduct pairwise comparisons of the elements at each hierarchy level to assess their relative importance.
4. **Priority Calculation:** Calculate the priority weights for each element using the eigenvalue method or other mathematical techniques.
5. **Consistency Check:** Assess the consistency of the judgments made during the pairwise comparisons to ensure reliability.
6. **Synthesis:** Aggregate the priority weights to determine the overall ranking of the alternatives.

By combining content analysis and AHP, this methodology provides a robust framework for understanding the complexities of the research topic, allowing for both rich qualitative insights and rigorous quantitative analysis.

3.2.1 Application of the AHP Model in This Study

In the first step, a hierarchical structure is created based on the information obtained from the content analysis model. The most important and prioritized stage in AHP is the establishment of the hierarchical structure. Information about how to construct the hierarchy is given in Fig. 2.

Figure 2 illustrates the various levels of the AHP process. The top level represents the decision problem's objective, followed by the criteria established in the second level. The third level consists of 18 sub-criteria, and the last level is the alternatives that will be prioritized based on these sub-criteria.

In the subsequent step, a comparative assessment of the criteria and alternatives within the hierarchical structure is conducted. To facilitate these comparisons, a scale featuring 17 bipolar and evenly spaced units is employed, resembling a semantic differential scale [47]. The units on this scale are described as "Intensity of Importance," detailed in Table 2.

In this study, for preference judgments, seven architectural experts with experience in both construction and traditional architecture were asked to perform pairwise comparisons between the criteria and sub-criteria. Table 3 illustrates how these judgments were made.

In scoring pairwise comparisons, the reciprocal condition is crucial: if the importance of element i relative to j is equal to k, then the importance of element j relative

Table 2 Intensity of importance table

Degree of importance	Definition	Explanation
1	Equal Importance	Both factors have the same importance
3	Moderate Importance	According to experience and judgment is more one factor important than the other
5	Strong Importance	One factor is strongly more important than the other
7	Very Strong Importance	One factor is strongly preferred at a higher level than the other
9	Extreme Importance	One of the factors is very important at a very high rate
2,4,6,8	Intermediate Values	These are the intermediate rates, they use when compromise is needed

Ref: [51]

Table 3 Comparative judgment structure of criteria

Criteria	Design	Construction	Performance
Design	1	5	8
Construction	1/5	1	4
Performance	1/8	1/4	1

Table 4 Normalized matrix and criteria weights

Criteria	Design	Construction	Performance	Criteria Weights
Design	0/233	0/676	0/615	0/508
Construction	0/349	0/135	0/308	0/264
Performance	0/419	0/189	0/077	0/228

to i will be equal to 1/k. This principle must be adhered to at all levels and pairwise comparison matrices within this model.

Using Expert Choice software, the criteria scores were obtained and shown in Table 3. Furthermore, a consistency test is needed to ensure that the consistency ratio is less than or equal to 0.10 to validate the judgment because this subjective process is based on judgments made by decision-makers. In this study, at each level of pairwise comparisons, the inconsistency rate was considered to be less than 0.1.

The normalized matrix for the criteria, the priorities vector, is given in Table 4.

At the third level, the pairwise comparison matrix of the sub-criteria is conducted, and the scores are calculated. Tables 5, 6, and 7 illustrate this process.

After determining the importance coefficients of the criteria and sub-criteria, the importance coefficient of the options must be established. At this stage, the preference of each option concerning each sub-criterion is evaluated. The basis for this evaluation is the same 9-point scale. After determining the importance coefficients for all sub-criteria concerning each option, the final score for each option will be determined by combining these importance coefficients while considering all scores across all hierarchical levels.

4 Discussion

"Yazd is a prime example of traditional architecture adapted to hot and arid climates worldwide. Beyond its aesthetic appeal, the architecture of Yazd is remarkably aligned with the region's climatic needs and conditions. The orientation of houses and specific building elements within the city's traditional fabric have contributed to the enduring resilience of these structures, despite the challenging climate [52].

Furthermore, studies of traditional architecture in historic houses of Yazd demonstrate a profound utilization of renewable, affordable, and clean energy sources throughout history, while minimizing reliance on non-renewable, polluting, and costly energy options. The region's architecture has been shaped by its geographic and ecological characteristics, as well as economic, cultural, and social factors—all of which are sustainability indicators [53]."

The traditional fabric of the world city of Yazd can be recognized as a climate-resilient zone. Over centuries, this fabric has been designed and constructed using indigenous knowledge and environmentally friendly materials, enabling it to withstand the desert's extreme heat and water scarcity. Climate-adaptive architecture in

Table 5 Comparative judgment structure of sub-criteria (design) and final score

Sub-Criteria (Design)	Site plan	Structural design	Architectural design	Cost plan	Building envelope	Ecological protection	Renewable energy	Material recycling	Final score
Site plan	1	5	4	1	2	1/2	1/4	1/2	0/164
Structural design		1	1	3	4	2	1	1	0/118
Architectural design			1	4	3	1	1	1	0/107
Cost plan				1	1/4	1/5	1/3	1/3	0/097
Building envelope					1	1/4	1/5	1/4	0/118
Ecological protection						1	1/2	1/2	0/130
Renewable energy							1	1	0/145
Material recycling								1	0/122

Table 6 Comparative judgment structure of sub-criteria (construction) and final score

Sub-Criteria (Construction)	Construction Safety	Air pollution	Noise pollution	Light pollution	Waste minimization	Final score
Construction Safety	1	4	4	4	1/3	0/293
Air pollution		1	1	2	1/2	0/145
Noise pollution			1	2	1/3	0/148
Light pollution				1	1/4	0/144
Waste minimization					1	0/269

Table 7 Comparative judgment structure of sub-criteria (performance) and final score

Sub-Criteria (Performance)	Living environment	Natural environment	Neighborhood	Thermal quality	Lighting quality	Final score
Living environment	1	3	4	1/3	1/2	0/192
Natural environment		1	5	4	4	0/287
Neighborhood			1	1/5	1/5	0/147
Thermal quality				1	1	0/198
Lighting quality					1	0/175

this city manifests itself as structures such as wind towers, domed roofs, mudbrick walls, and central courtyards.

Based on this, in this research, four traditional houses in the city of Yazd have been considered, and the elements related to Healthy Buildings have been examined within them.

Table 8 presents the architectural patterns of Yazdi houses, which are over 100 years old and remain in use today. While some have undergone wear and tear, most have been renovated and adapted for new purposes. The sustainability of these buildings reflects their strong resilience despite climatic challenges. The typology of traditional Yazdi houses is characterized by their central two-story organization:

a. Houses with a single central courtyard.
b. Houses with multiple internal courtyards.

Within the category of houses with a single courtyard, further subcategories exist based on the number of rooms along the southern side of the courtyard. This central courtyard, a feature prevalent in Iranian architecture for approximately 8000 years [54], promotes an introverted spatial organization and results in houses that are expansive, voluminous, and offer panoramic views [55].

Table 8 Samples of traditional healthy buildings in Yazd

Name of house	Age	Type of Courtyards	Repaired	Lived & usable	Performance (after repairing)	Plan	Perspective	Picture
Akhavansegari house	Qajar and Pahlavi era	Inner and Outer courtyard	Yes	Yes	As a student dormitory	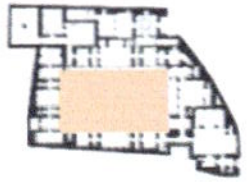		
Arab kermani house	170 years ago	Inner and Outer courtyard	No	Yes	As a house			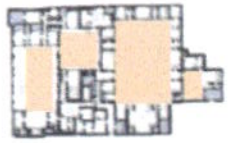
Rasoolian house	120 years ago	Inner and Outer courtyard	Yes	Yes	As an architecture college		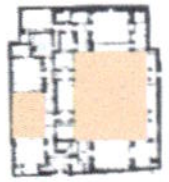	
Arabha house	150 years ago	Inner and Outer courtyard	Yes	Yes	As an office	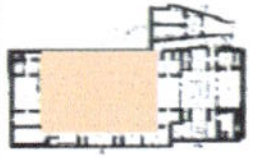		

Source Authors

4.1 Examination of Healthy Building Criteria in Traditional Houses of Yazd

Given the pressing climate crisis, it is now more crucial than ever to consider climatic conditions in the design and construction of human settlements. This issue is particularly evident in desert cities that grapple with water scarcity and harsh climatic conditions. Considering the antiquity of Yazd and the durability of its traditional buildings, it seems that these houses can exemplify the concept of climate resilience. Moreover, the architectural elements of these structures were designed with a specific philosophy that ensured the physical and mental well-being of their inhabitants. The primary architectural elements of these houses are depicted in Fig. 1, aligning with the theoretical framework of the research.

The main physical components of the homes consist of:

- **Courtyard:** Historical buildings in Yazd exhibit a centralized organizational structure centered around a courtyard. This creates an introverted layout where all living and other spaces are arranged around one or more central courtyards. The courtyard significantly influences the microclimate of these houses (Fig. 4).

  The air within the courtyard is relatively still but is constantly replenished with fresh outdoor air. This process allows lighter vapors or gases (like H_2O) to rise, while heavier gases (like CO2) tend to accumulate at the courtyard's bottom [56]. Introducing fresh outdoor air into buildings is crucial for maintaining healthy indoor environments (Fig. 5).
- **Badgir (Wind-Catcher):** The badgir is a fundamental component of natural ventilation and air circulation (Fig. 4). Its primary function is to naturally cool building interiors without energy consumption in the hot, dry climate of Yazd. Badgirs effectively reduce heat and regulate indoor temperatures, creating comfortable living spaces through natural convection and clean air ventilation [57].
- **Sardab, Payab, and Qanat**: In some houses, there are Sardabs, Qanats, and Payabs connected to the Badgir. The wind circulates within the Badgir and is directed toward the water (Fig. 5). Water flows through the Qanat, which is located within the Sardab. The Payab is the main entrance to the Sardab or Qanat. Family members can relax in the cool, wet, and pleasant Sardab during the hot summer without relying on mechanical cooling systems. Additionally, the Qanat, a sustainable water supply system [58], plays a beneficial role in the health of Yazd's buildings (Fig. 7).
- **Panj-Dari:** The Panj-Dari was the heart of daily life in traditional Yazdi homes, a spacious room adjacent to the main family living area. Often connected to a large balcony overlooking the central courtyard through five expansive windows, the Panj-Dari was a focal point of the house. Five glass-paneled wooden doors, separated by sturdy stone columns, opened onto the courtyard, creating a visually striking symmetrical facade (Fig. 6). Smaller rooms, such as the Se-Dari, were dedicated to family living, while larger spaces like the Talar (hall) and Panj-Dari, positioned centrally, served as reception areas. The Panj-Dari was remarkably

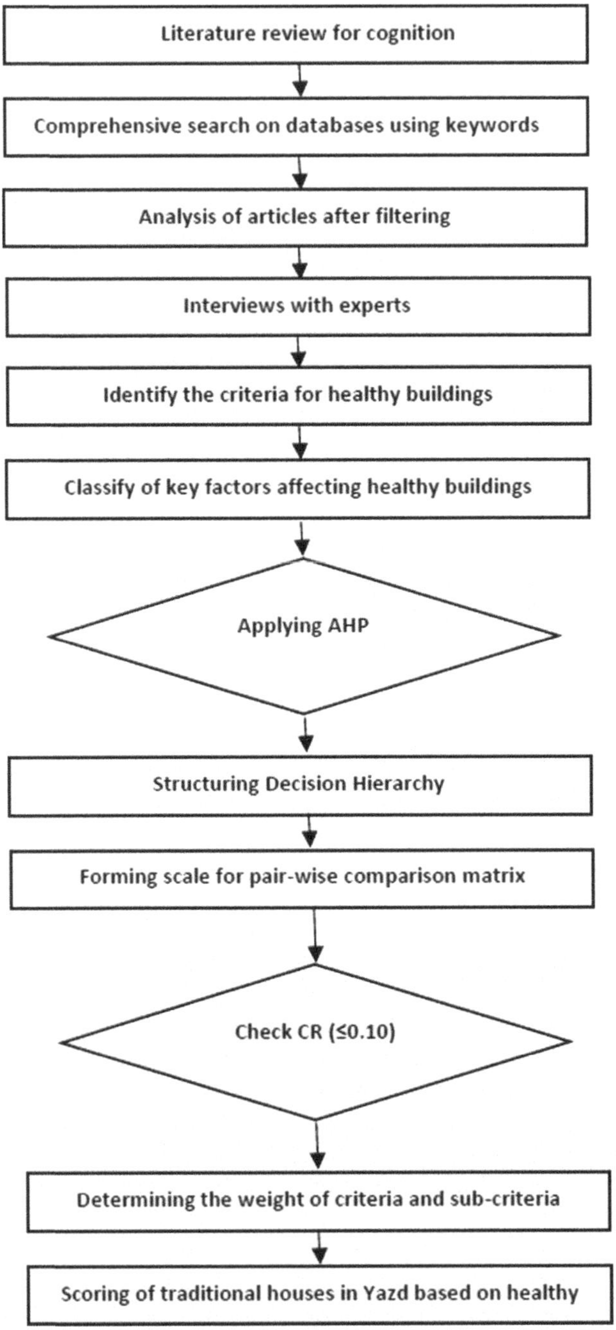

Fig. 1 Methodology flow chart

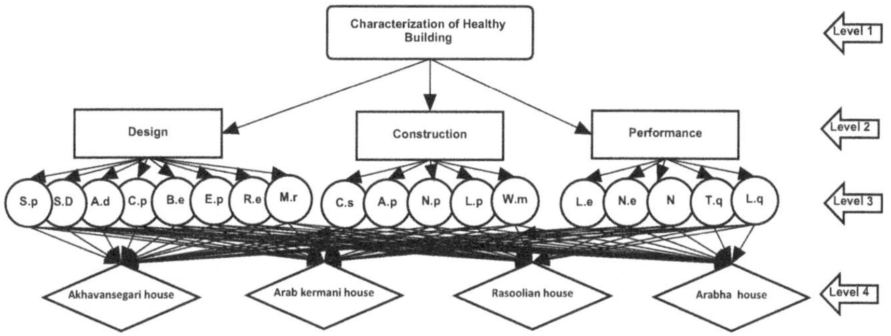

Fig. 2 The Hierarchical Structure of AHP

esteemed as a guest room [59]. In grander residences, this space often evolved into the formal Talar. All these rooms, aligned with the courtyard's axis, featured elegant wooden doors adorned with colorful Orossi glass panes, enhancing the interior's aesthetic appeal (Fig. 8).

The remaining elements in Fig. 3 have more significant psychological impact and involve Sky-scape, Landscape, Soundscape, Spirituality, and Aesthetics, as illustrated in Figure.

Given the existing evidence, the advantages of healthy architectural elements like wind-catchers and central courtyards in traditional Yazdi homes can be tabulated as 3. This Table 9, while respecting the traditional architectural heritage of Yazd, clearly indicates the positive effects of these elements on the well-being of inhabitants, both physically and mentally.

4.2 Pairwise Comparison of Alternatives for Ranking Traditional Yazd Homes

Based on the application of the AHP model, the results of the decision-makers extracted from each of the seven experts and the preference judgments, all having a CR less than or equal to 0.1, indicate that among the criteria considered regarding healthy buildings, the Design criterion with a weight of 0.508 has been prioritized by the experts over the other two criteria, namely Construction and Performance. Additionally, the sub-criteria weights have varied among the experts, as shown in Fig. 9.

As shown in Fig. 9, the sub-criteria of site plan, construction safety, and natural environment were deemed most important within their respective categories of Design, Construction, and Performance.

To assess the health criteria of four traditional Yazd houses, pairwise comparison matrices were constructed for each sub-criterion based on expert opinions. The final weights of these houses are shown in Table 10.

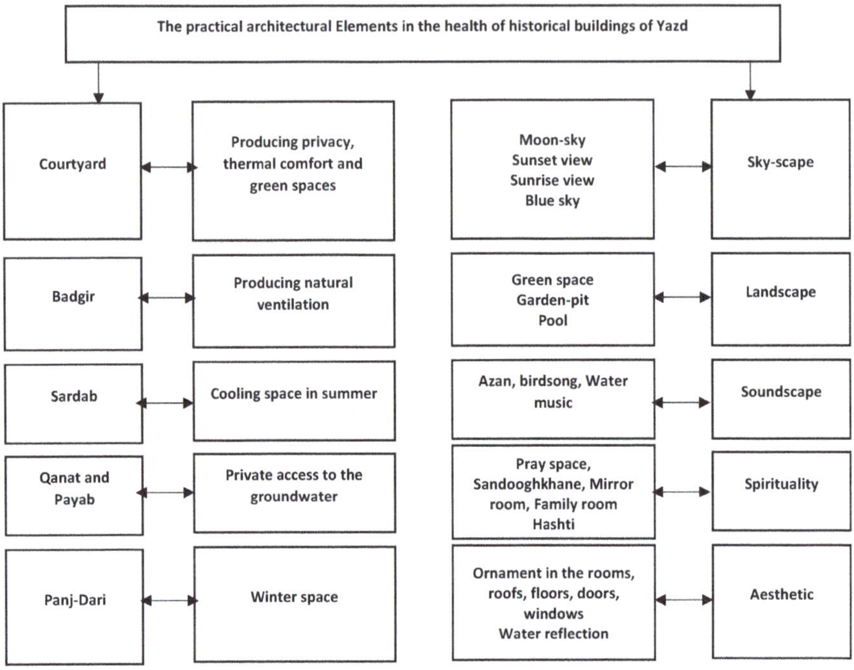

Fig. 3 The built elements of healthy buildings in historical housings in Yazd

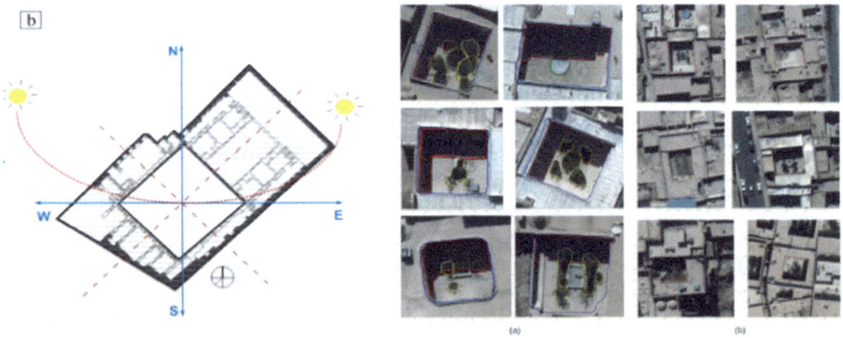

Fig. 4 Courtyard influence on the microclimate of old Yazd Houses

Up to this point, the importance coefficients of the criteria and sub-criteria concerning the study's objective and the importance coefficients (scores) of the options concerning each sub-criteria have been determined. Finally, to determine the final score of each alternative (traditional houses of Yazd), the coefficients, as mentioned earlier, have been combined, and the result is shown in Fig. 10. Accordingly, although all the considered options have similar scores, the Rasoolian house

Fig. 5 Performance mechanism of courtyard buildings in Yazd

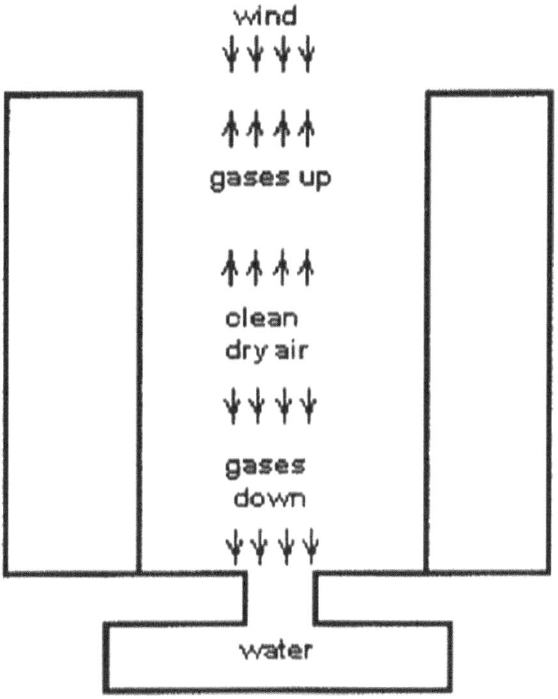

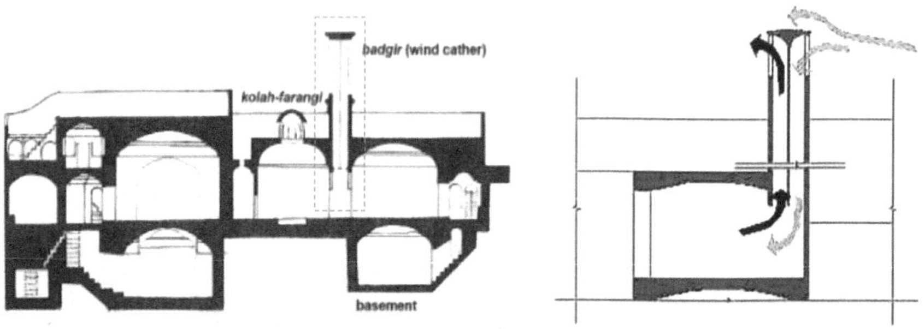

Fig. 6 Badgir in Yazd *Source* Adapted from [60, 61]

has a higher score compared to the other options, indicating that the main criteria for a healthy building are present in it.

Fig. 7 Sardab and Payab; the basement plan and section drawings are on different scales

Fig. 8 Panj_Dari in two historical houses of Yazd

Table 9 The benefits of Healthy Buildings in the traditional architecture of Yazd

Futures of healthy buildings	Strategies for healthy buildings	Healthy building criteria
God; Allah is present in the housing and house planning	Sandoghkhaneh, Pray space	Mentally Healthy
Spirituality spaces, not soulless spaces	Green space, Ornamented rooms, Family room, Neighboring space	
An intimated family atmosphere: The family is not separate in the house	Inner courtyard, Panj-Dari	
An intimated neighboring atmosphere: There is neighborhood (get-together) space	Hashti as an entrance	
All scape is pleasurable	Landscape, soundscape, sky-scape, starscape, moonscape, sunset and sunrise views (courtyard and top roof)	
Increase Privacy	Hashti, Dalan, Inner and Outer courtyard, gates, two separate door knockers for men and women	
Minimizing Energy Needs	Optimal shading in summer and winter; Natural heating; Greatest solar gain in winter; Best natural ventilation and passive cooling; natural light (daylight and moonlight)	Physically Healthy
Passive Water Supply System	Use passive water supply systems and minimize the energy needed to provide water	
'Green' Energy Sources	Sun, Wind, Groundwater, and Qanqt, Soil	
Effective green area	Location of vegetation in the house: garden-pit	
Plentiful Material	Local materials: adobe, mud, mud brick	
Low Emitting Materials	Earthen construction materials: clay, cob, adobe, and unfired clay bricks	
Plentiful structure	Local construction techniques: Arch and dome technique with bearing wall; masonry buildings	
Comfort	Access to fresh air; Purifying air with plants; lifetime homes; Thermal insulation; Sound insulation; Humid air in the Sardab and Courtyard	

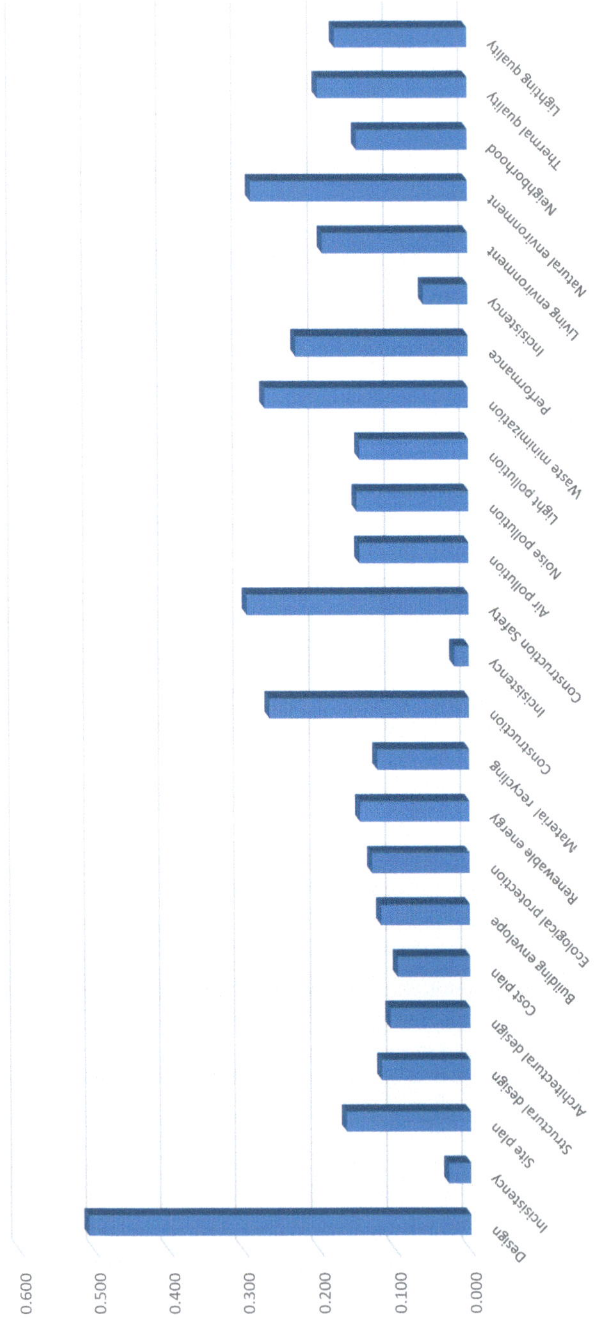

Fig. 9 Ranking of the main criteria and sub-criteria considering the goal

Table 10 Examination of Healthy Building factors in case studies

Name of building	Design							
	Site plan	Structural design	Architectural design	Cost plan	Building envelope	Ecological protection	Renewable energy	Material recycling
Akhavansegari house	0/28	0/25	0/26	0/21	0/30	0/26	0/25	0/25
Arab kermani house	0/27	0/23	0/23	0/17	0/13	0/24	0/24	0/24
Rasoolian house	0/18	0/27	0/26	0/46	0/30	0/25	0/26	0/26
Arabha house	0/27	0/25	0/26	0/17	0/28	0/25	0/25	0/25

Name of building	Construction					Performance						
	Construction Safety	Air pollution	Noise pollution	Light pollution	Waste minimization	Living environment	Natural environment	Neighborhood	Thermal quality	Lighting quality	Acoustic quality	Air quality
Akbavansegari house	0/15	0/27	0/27	0/27	0/29	0/26	0/28	0/24	0/25	0/25	0/25	0/25
Arab kermani house	0/28	0/24	0/26	0/21	0/31	0/24	0/13	0/25	0/25	0/25	0/25	0/24
Rasoolian house	0/43	0/24	0/22	0/29	0/18	0/26	0/35	0/24	0/25	0/26	0/25	0/26
Arabha house	0/13	0/24	0/24	0/23	0/22	0/24	0/25	0/26	0/26	0/24	0/25	0/24

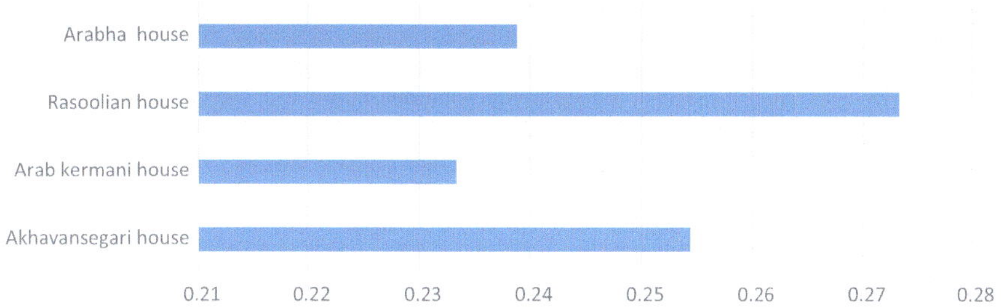

Fig. 10 The final prioritization of traditional houses in Yazd is based on healthy building criteria

5 Conclusions

The findings of this study emphasize the pivotal role of traditional architecture in addressing contemporary sustainability challenges. By examining the case of Yazd, a city renowned for its rich architectural heritage and challenging climate, this research has demonstrated the potential of traditional building practices to inform the design of modern, healthy, and resilient structures. The results suggest a strong correlation between the principles embodied in traditional Yazd houses and emerging standards for sustainable and healthy buildings.

Through a comprehensive analysis of 18 factors across design, construction, and performance, it was found that most traditional Yazd houses align with the standards of healthy buildings. This indicates that indigenous architectural principles in adverse climates are well-suited to healthy building criteria and that the architectural model of traditional Yazd houses can serve as a valuable blueprint for modern housing.

While this research offers valuable insights, further investigations are necessary to fully understand the complex interplay between traditional building practices and contemporary sustainability challenges. In-depth studies on specific building components, energy performance, and occupant comfort in traditional Yazd houses can provide more granular data for future design guidelines. Additionally, exploring the potential for replicating the success of Yazd's architectural heritage in other regions with similar climatic conditions could contribute to the global discourse on sustainable urban development.

Ultimately, this research highlights the importance of preserving and revitalizing traditional building knowledge to create sustainable and resilient cities. By recognizing the value of architectural heritage, we can develop innovative solutions that address the pressing challenges of our time while respecting the wisdom of the past.

Acknowledgements On behalf of my team, I would like to thank the cultural heritage experts of Yazd for their generous collaboration on this project.

References

1. Nicolini E (2024) Climate change adaptation and mitigation and historic centers preservation. Underway and repeatable technological design solutions. Cities 152, 105174
2. Holubchak K, Obynochna ZV (2022) Architectural & urban planning strategy "sponge city" as a means of fighting the consequences of global climate change. Містобудування та територіальне планування. https://doi.org/10.32347/2076-815x.2022.80.149-158
3. Al-Sayed MM, Alanizi A (2023) Resilient cities to mitigate effects of climate change. In IOP Conference Series: Earth and Environmental Science, (Vol. 1129, No. 1, p. 012029). IOP Publishing
4. Bouramdane AA (2023) Shaping resilient buildings and cities: climate change impacts, metrics, and strategies for mitigation and adaptation. Inf Syst Smart City 3(1). Buildings 2023, Buildings Performance Institute Europe. Building 4 People: People-Centric Buildings for European Citizens
5. Ye X, Niyogi D (2022) Resilience of human settlements to climate change needs the convergence of urban planning and urban climate science. Comput Urban Sci 2(1):6. Available online: https://bpie.eu/wp-content/uploads/2017/12/White-Paper2017.pdf. Accessed on 1 March 2022
6. Marschütz B, Bremer S, Runhaar H, Hegger D, Mees H, Vervoort J, Wardekker A (2020) Local narratives of change as an entry point for building urban climate resilience. Clim Risk Manag 28:100223
7. World Health Organization (WHO) (2018) The Constitution of the World Health Organization
8. Samet JM, Spengler JD (2003) Indoor environments and health: moving into the 21st century. Am J Public Health 93(9):1489–1493
9. Bernstein ES, Turban S (2018) The impact of the 'open ' workspace on human collaboration. Philos Trans Roy Soc B Biol Sci 373(1753):20170239
10. Li Y, Rong Y, Ahmad UM, Wang X, Zuo J, Mao G (2021) A comprehensive review on green buildings research: bibliometric analysis during 1998–2018. Environ Sci Pollut Res 28:46196–46214
11. Loftness V, Hakkinen B, Adan O, Nevalainen A (2007) Elements that contribute to healthy building design. Environ Health Perspect 115(6):965–970
12. Elena C, Fabio F (2017) Energy resilience of historical urban districts: a state of art review towards a new approach. Energy Procedia 111:426–434
13. Jigyasu R (2019) Managing cultural heritage in the face of climate change. J Int Aff 73(1):87–100
14. Teba T, Elzoghby NM, Gharib N (2022) The role of architectural heritage in offering a more resilient lockdown in Egypt. In International Conference on "Health & Environmental Resilience and Livability in Cities-The challenge of climate change, Springer Nature Switzerland, Cham, pp 65–91
15. Kusumaningrum AE, Hieu LHT, Putra S (2021) Implementation of reliability buildings based on health principles. In AIP Conference Proceedings, vol. 2406. AIP Publishing.

16. El-Daghar K (2022) Reduce environmental impacts on architectural heritage—case study six sigma technique to achieve high-performance for historic buildings in Lebanon. Conservation of Architectural Heritage (CAH) Embodiment of Identity. Springer International Publishing, Cham, pp 301–318
17. Fuentes B (2022) Healthy buildings: a combination of nature and technology can secure a sustainable future. In 9th International Conference on Harmonisation between Architecture and Nature, ARC 2022, pp 237–248
18. Hu M, Hu M (2021) A brief history of health and the built environment. Smart technologies and design for healthy built environments, Springer, Cham, pp 1–19
19. Architectural Society of China (2017) Healthy building evaluation standard T/ASC 02–2016. China Construction Industry Press, Beijing, China
20. Carmichael L, Prestwood E, Marsh R, Ige J, Williams B, Pilkington P, Michalec A et al (2020) Healthy buildings for a healthy city: is the public health evidence base informing current building policies? Sci Total Environ 719:137146
21. Lin Y, Yuan X, Yang W, Hao X, Li C (2022) A review on research and development of healthy building in China. Buildings 12(3):376
22. Felicioni L, Lupíšek A, Gaspari J (2023) Exploring the common ground of sustainability and resilience in the building sector: a systematic literature review and analysis of building rating systems. Sustainability 15(1):884
23. Grierson D (2003) Arcology and Arcosanti: towards a sustainable built environment. Electron Green J 1(18)
24. Mao P, Qi J, Tan Y, Li J (2017) An examination of factors affecting healthy building: an empirical study in east China. J Clean Prod 162:1266–1274
25. Gowri K (2004) Green building rating systems: An overview. ASHRAE J 46(11):56
26. Yudelson J (2008) Green building through integrated design. Green building through integrated design. McGraw-Hill, 125e125
27. Castleton HF, Stovin V, Beck SB, Davison JB (2010) Green roofs; building energy savings and the potential for retrofit. Energy Build 42(10):1582–1591
28. Kats G (2003) Green building Costs and financial benefits. Massachusetts Technology Collaborative, Boston, MA
29. Pan Y, Yin R, Huang Z (2008) Energy modeling of two office buildings with data center for green building design. Energy Build 40(7):1145–1152
30. Zhang X, Wu Y, Shen L, Skitmore M (2014) A prototype system dynamic model for assessing the sustainability of construction projects. Int J Project Manage 32(1):66–76
31. Stafford A, Gorse C, Shao L (2011) The retrofit challenge: delivering low carbon buildings. The Centre for Low Carbon Solutions, York
32. Chen JH, Zheng BH, Yan YP (2012) A prospective study of blood pressure dynamic change and related factors in construction workers in high-altitude regions. Med J Chin PAPF 23(7):583–585
33. Glaeser EL, Kahn ME (2010) The greenness of cities: Carbon dioxide emissions and urban development. J Urban Econ 67(3):404–418
34. Adedeji YMD, Taiwo AA, Olotuah OA, Fadairo G, Ayeni DA (2013) Low carbon construction materials and techniques for sustainable housing development in Nigeria. In AEI 2013: Building Solutions for Architectural Engineering, pp 643–651
35. Chau CK, Leung TM, Ng WY (2015) A review on life cycle assessment, life cycle energy assessment, and life cycle carbon emissions assessment on buildings. Appl Energy 143:395–413
36. Bungau CC, Bungau T, Prada IF, Prada MF (2022) Green buildings as a necessity for sustainable environment development: dilemmas and challenges. Sustainability 14(20):13121
37. Kibert CJ (2016) Sustainable construction: green building design and delivery. John Wiley & Sons
38. Baharetha SM, Al-Hammad AA, Alshuwaikhat HM (2013) Towards a unified set of sustainable building materials criteria. In ICSDEC 2012: Developing the frontier of sustainable design, engineering, and construction, pp 732–740

39. Franco MAJQ, Pawar P, Wu X (2021) Green building policies in cities: a comparative assessment and analysis. Energy Build 231:110561
40. Wang C, Zhan J, Xin Z (2020) Comparative analysis of urban ecological management models incorporating low-carbon transformation. Technol Forecast Soc Chang 159:120190
41. Cioruța B, Coman M (2018) Beyond the construction, design, and planning scenarios of eco-buildings. Asian J Adv Res R 2(2):1–13
42. Li F, Wang R, Paulussen J, Liu X (2005) Comprehensive concept planning of urban greening based on ecological principles: a case study in Beijing. China. Landscape Urban Plann 72(4):325–336
43. Haider S, Ahmad J, Ahmed M (2018) Identifying barriers to implementation of health promoting schools in Pakistan: the use of qualitative content analysis and fuzzy analytic hierarchy process. Int J Adv Appl Sci 5(4):56
44. Eslami Mahmoodabadi S, Mofidi Shemirani SM, Habib F (2021) Investigating the principles of indigenous housing architecture of hot and dry climate of Iran with the formal impact approach of the building in promoting thermal comfort of building users. Islamic Art Stud 17(40):21–39
45. The WELL Building Standard-V2 Pilot; International WELL Building Institute (IWBI): New York, NY, USA, 2016.
46. Iddi S, Muindi K, Gitau H, Mberu B (2022) Characterization of healthy housing in Africa: method, profiles, and determinants. J Urban Health 99(1):146–163
47. Şahin M, Yurdugül H (2018) A content analysis study on the use of analytic hierarchy process in educational studies. J Meas Eval Educ Psychol 9(4):376–392
48. Anuar NB, Papadaki M, Furnell S, Clarke N (2013) Incident prioritisation using analytic hierarchy process (AHP): Risk Index Model (RIM). Secur Commun Netw 6(9):1087–1116
49. Borade AB, Kannan G, Bansod SV (2013) Analytical hierarchy process-based framework for VMI adoption. Int J Prod Res 51(4):963–978
50. Mushtaha E, Alsyouf I, Al Labadi L, Hamad R, Khatib N, Al Mutawa M (2020) Application of AHP and a mathematical index to estimate livability in tourist districts: the case of Al Qasba in Sharjah. Front Architectural Res 9(4):872–889
51. Saaty TL (1990) How to make a decision: the analytic hierarchy process. Eur J Oper Res 48:9–26. http://dx.doi.org/10.1016/0377-2217(90)90057-I
52. Jamalpour S, Arbaban Esfahani A (2014) The impact of climate on the formation of Yazd's house architecture. National Conference on Iranian Indigenous Architecture and Urbanism, SID. https://sid.ir/paper/826666/fa
53. Borjyan F, Kazemian F (2012) A sustainable development perspective on the architecture of different types of historic houses in Yazd. First National Conference on Architecture, Restoration, Urban Planning, and Sustainable Environment, Hamedan. https://civilica.com/doc/263282
54. Memarian MH (1996) Traditional Iranian architecture: a historical analysis and typology. Sarcheshmeh Publications, Tehran
55. Soltani M (2017) Yazd: traditional houses and the transformation of urban space. Springer, Cham
56. Chen Q, Clements-Croome D (2020) Healthy buildings: lessons from ancient China
57. Mikaeili M, Memlük Y (2012) Badgir: design with nature, a traditional architectural and climate element in hot dry region of Iran. In ICONARCH International Congress of Architecture and Planning, pp 136–146
58. Nasiri F, Mafakheri MS (2015) Qanat water supply systems: a revisit of sustainability perspectives. Environ Syst Res 4:1–5
59. Einifar A, Abroon N (2021) Explaining the continuity of hospitality from Iranian houses to contemporary apartments. Armanshahr Archit Urban Dev 13(33):167–179
60. Abouei R (2006) Conservation of Badgirs and Qanats in Yazd, Central Iran. In The 23rd Conference en passive and low Energy Architecture, Geneve, Switzerland, pp 6–8
61. A'zami A (2005) Badgir in traditional Iranian architecture. In International conference Passive and low energy cooling for the built environment, pp 1021–1026

Vulnerability of Urban Housing to Climate Change in Cameroon

Suiven John Paul Tume, Yongka Bernard Nyuyse, Clarius Kuyar, Ngoran Larissa Tomla, Ngoran Hassan Afoni, Foka Sherifatu Kinyuy, and Ngala Elvis Ngum

Abstract Climate change impacts on human settlements and community well-being is a significant concern, especially in vulnerable countries like Cameroon, which has a wide range of ecological zones, including Coastal Lowlands, Equatorial Rainforest, Western Highlands, Sudan-Savannah, and the Sahel. The objectives of this study are (1) evaluate changing climatic conditions in Cameroon's ecological zones; (2) analyse the vulnerability of urban residential housing to changing climatic conditions; (3) propose adaptation strategies and sustainable architectural designs to mitigate the impact of climate change, (4) assess policy implications and stakeholder engagement for sustainable urban housing development. A literature review was conducted to achieve these objectives on the nexus between climate and human settlements. Meteorological data (rainfall) were also analysed and 200 structured questionnaires were administered to municipal authorities to assess the vulnerability and adaptation of urban residential housing. Findings revealed that the rapidly urbanising coastal cities of Douala, Limbe, and Kribi are experiencing extreme precipitation, flooding, and sea level rise. Similarly, the capital city, Yaounde, is facing severe rainfall, leading to floods and landslides. In the Western Highlands, human settlements in Bafoussam and Bamenda are at risk of landslides and runoff. In the northern Sudano-Sahelian region, extreme heat and desertification pose significant threats to residential housing development. Nature-based Climate Solutions (NbS) such as green corridors are recommended to absorb and filter stormwater to promote climate-resilient communities. Developing climate-resilient residential housing in Cameroonian cities requires tailored community-level interventions for each ecological zone.

Keywords Ecological design · Green urbanisation · Nature-based solutions · Resilience · Sustainable architecture

---

S. J. P. Tume (✉) · Y. B. Nyuyse · C. Kuyar · N. L. Tomla · N. E. Ngum
Department of Geography and Planning, The University of Bamenda, Bambili, Cameroon
e-mail: suiven.tume@outlook.com

N. H. Afoni · F. S. Kinyuy
Department of Geography, University of Buea, Buea, Cameroon

A. Cheshmehzangi et al. (eds.), *Designing Healthy Buildings and Communities*, Urban Sustainability, https://doi.org/10.1007/978-981-96-3057-8_4

1 Introduction

Vulnerability of urban areas to climate change is a global concern, especially in rapidly urbanizing areas and informal settlements [1–4]. As cities expand and populations grow, the effects of climate-related shocks become more significant, impacting infrastructure, livelihoods, and human well-being. For example, the urban heat island effect has been identified as a significant climate-related phenomenon affecting cities, worsening temperature extremes and associated health risks [5]. Various factors contribute to the vulnerability of urban housing, including building characteristics, spatial organisation, and industrial activities [5]. These factors interact with climate stressors, creating complex vulnerability patterns that differ across urban areas. Researchers have used tools like the Livelihood Vulnerability Index of the IPCC [6] to better understand and measure this vulnerability. Such assessments have revealed that peri-urban and poor households are often the most susceptible to climate change impacts, with noticeable gender disparities in adaptation capabilities [6].

In response to these challenges, a new vulnerability paradigm is emerging as a tool for city decision-makers to analyze adaptation strategies [7]. This approach provides a clearer understanding of urban vulnerability, enabling targeted and effective adaptation measures. Integrated approaches are increasingly recognized as essential for enhancing urban resilience, including investments in peri-urban areas, poverty reduction initiatives [6]. Urban climate vulnerability is pronounced in developing countries, where rapid urbanisation often outpaces infrastructure development and adaptive capacity [8]. African cities are among the least studied regarding climate change impacts, despite being highly exposed to natural hazards [9]. This knowledge gap underscores the need for increased research attention and targeted interventions in African urban centres.

Urban areas in Cameroon are facing increasing risks from climate change impacts, particularly flooding, due to rapid and often unplanned urbanization [10]. The vulnerability is even higher in slum areas, which account for an estimated 65–70% of urban areas in the country [8]. Other contributing factors to the vulnerability of Cameroonian cities include rugged or flat topography, poor infrastructure, and inadequate waste management [8]. Climate variability in Cameroon is evident in heavy rainfall events, temperature extremes, and violent winds [11]. These climate stressors, combined with local vulnerabilities, pose significant risks for urban populations. Flood vulnerability in northern Cameroon is further worsened by ineffective institutional performance and weak legislative frameworks [12]. Yaoundé provides a clear example of increasing urban vulnerability, as it has experienced heightened flood vulnerability over the past 15 years due to rainfall surpluses and deficits, demographic pressure, and uncontrolled urbanisation [11]. This situation illustrates the complex interplay between climate change, urban development, and socio-economic factors in shaping urban vulnerability.

This chapter attempts to bridge some empirical gaps in studies on the effects of climate change in urban areas in Cameroon that have often focused more on flooding and the disruption of urban traffic flow, without considering the vulnerability of housing like [11–16] who revealed that surging flood frequency in Cameroon in the last three decades has been characterized by independent, isolated cases, limiting the understanding of possible correlations across geographical space and time. The objectives of this chapter, therefore are: (1) to assess changing climatic conditions in Cameroon's ecological zones, (2) to analyse the vulnerability of urban residential housing to changing climatic conditions; (3) to propose adaptation strategies and sustainable architectural designs to mitigate the impacts of climate on urban housing, (4) assess policy implications and stakeholder engagement for sustainable urban housing development.

2 Methods of Study

Cameroon is often described as a babble of tongues, and Africa in miniature. Located in the Gulf of Guinea (Fig. 1a and b), the country has five agroecological zones (Sudano-Sahelian, Guinea High Savannah, Western Highlands, Mono-Modal Rainforest, and Bi-Modal Rainforest) [17, 18]. The country also has ten administrative regions (Adamawa, Centre, East, Littoral, Far North, North, Northwest, South, Southwest, and West), with an estimated population of about 29 million inhabitants (2024 United Nations estimates). Cameroon spans an area of 475,650 km^2, with a north-south distance of 1,500 km (2–13°N) and an east-west distance of 800 km (9–16°E). The country has experienced an annual population growth rate of 2.5%, which has reached 4.3% in urban areas. Notably, uncontrolled urbanisation has seen the urbanisation rate rise from 52% in 2010 to 57% in 2019 [19].

Cameroon's varied latitude has resulted in a range of rainfall patterns, from a monomodal rainfall deficit in the Sahelian agro-ecological zone (500 to 800 mm) to a monomodal rainfall (1800 to 2800 mm) in the high savannah and highland agro-ecological zone, and relatively abundant bimodal rainfall (1500–2000 mm) in the forest agroecological zone [19]. The coastal agroecological zone experiences significant monomodal rainfall (3000–8000 mm). Temperature also varies across different regions, with a range of 20 °C to 35 °C and variations of 3 °C to more than 12 °C [19].

Cameroon is home to six main types of ecosystems, each with a diverse range of agro-pastoral production systems [19, 20]. From steppe and Yaérés in the Far North to savannah in the North, Adamawa, and the Western Highlands, as well as semi-deciduous forests in the Centre-South, and evergreen forests and mangroves in the coastal zone, the country boasts a rich flora and fauna. Again, Cameroon ranks fourth in terms of flora richness and fifth in terms of faunal diversity in Africa, with 8300 species of plants and 335 species of mammals. However, this rich biodiversity is under threat from illegal and anarchic logging, wildlife and mining, uncontrolled land use for slash-and-burn agriculture, and unsustainable development of agro-industries. The estimated annual net deforestation rate of 0.6% coupled with a low rate of reforestation (0.1%) indicates an ongoing loss of biodiversity through deforestation. Deforestation does not only contribute to climate change but also affects the

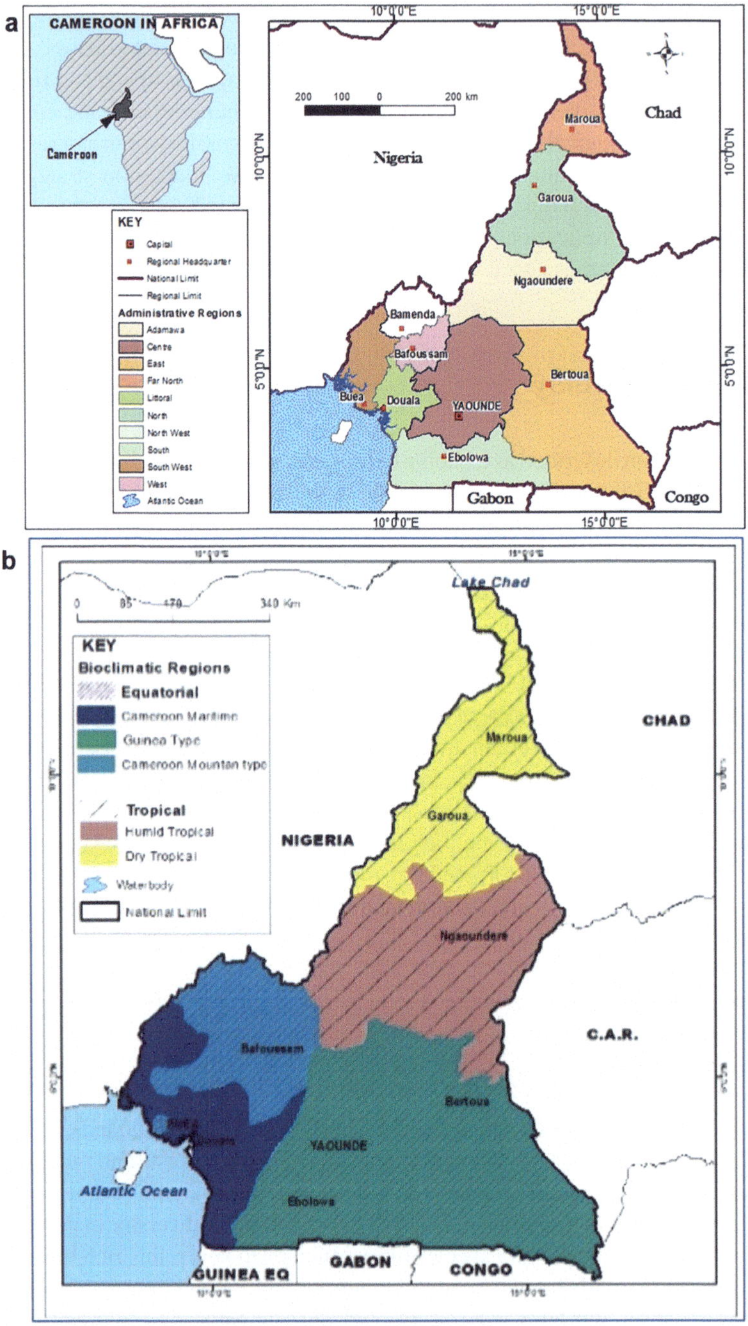

Fig. 1 Layout of and ecological zones of Cameroon

vulnerability of local and indigenous populations. Cameroon's large forest area is increasingly degraded by agro-pastoral activities, mining, and related development projects. Moreover, the country is experiencing significant population growth, with an estimated 29 million inhabitants in 2024 (United Nations estimates), resulting in an average population density of 56 inhabitants/km^2, which varies from 7 to 200 inhabitants/km^2 across the country [19].

2.1 Data Collection and Presentation

A mixed-methods research design, integrating qualitative and quantitative approaches was employed in this study. A questionnaire was administered to municipal authorities in Douala, Limbe, Kribi, Yaounde, Buea, Bertoua, Ebolowa, Bamenda, Bafoussam, Ngaoundere, Garoua and Maroua (Table 1). Quantitative methods were employed through the vulnerability of urban housing to climate change and environmental conditions, impacts, and the capacity for autonomous adaptation. Quantitative aspects involved climate risk analysis, considering factors beyond climate change, such as population vulnerability, land use, income levels and institutional perceptions [19]. Data collection for the study was multifaceted. It involved stakeholder engagement, where researchers interacted with local governments, communities, and experts to gather insights on climate change trends and ecological vulnerability, socio-economic vulnerability and urban housing, adaptation strategies and sustainable architectural design, policy implications, and stakeholder engagement [19]. Responses captured from this survey were treated in the Statistical Package for Social Sciences (SPSS), Version 25, and presented in graphs and tables. The main climatic element used in this study is rainfall, where mean monthly data was collected from the National Meteorological Service from 1981 to 2023 for all the cities implicated in this study [13]. The data was presented in time series graphs to show trends.

City maps were generated using ArcGIS 10.8, incorporating satellite images from the Google Earth base map to access real-time images of the specified areas. The study focused on the regions of Yaounde, Douala, Bamenda, Bafoussam, Garoua, Ngaoundere, Maroua, Bertoua, Buea, Ebolowa, Limbe, and Kribi. The shapefiles representing the boundaries of these locations were derived from the administrative map of Cameroon and arranged in layout mode to create the final maps for each area. These maps were configured to the WGS 1984 coordinate system.

Table 1 Distribution of questionnaires

Ecological zone	Selected cities	Population[a]	% of national population	Questionnaires administered	Questionnaires retrieved	% retrieved
Coastal Lowlands	DLA	1,338,082	27.93	40	34	14.05
	LBE	72,106	1.51	13	10	4.13
	KRB	55,224	1.15	10	8	3.31
Forest Zone	YDE	1,299,369	27.12	34	31	12.81
	BUA	47,300	0.99	13	11	4.55
	BER	218,111	4.55	11	9	3.72
	EBO	87,875	1.83	15	11	4.55
Western Highlands	BDA	393,835	8.22	25	22	9.09
	BAF	290,768	6.07	17	14	5.79
Sudan Savannah	NGD	231,357	4.83	15	13	5.37
Sahel	GAR	436,899	9.12	25	19	7.85
	MAR	319,941	6.68	24	18	7.44
Total		4,790,867	100	242	200	82.66

DLA: Douala, LBE: Limbe, KRB: Kribi, YDE: Yaounde, BUA: Buea, BER: Bertoua, EBO: Ebolowa, BDA: Bamenda, BAF: Bafoussam, NGD: Ngaoundere, GAR: Garoua, MAR: Maroua
[a]https://www.worldometers.info/world-population/cameroon-population/

3 Results

3.1 Climate Change Trends and Ecological Vulnerability

Cameroon like other tropical countries has witnessed changes in climatic conditions, with severe consequences in rural and urban areas. Most of these conditions tend to be hazardous for urban housing and development, thereby increasing vulnerability and other socio-economic and environmental shocks. Climatic elements tend to be hazardous when they occur in extremes. In this study, the sampled population indicated that increasing temperatures and rainfall, as well as increasing temperatures and decreasing temperatures are recurrent in all the ecological zones of Cameroon (Fig. 2). Increasing temperature and decreasing rainfall in more recurrent in all the ecological zones than increasing temperatures and rainfall. Despite the general decrease in rainfall, extreme rainfall events frequently cause flooding all over the country. Different parts of the country exhibit varied monthly rainfall patterns, with varied local effects.

From the onset of the rainy season, the recurrence of flooding and associated effects is recorded all over the national territory. There are different patterns in mean monthly rainfall in Cameroon (Fig. 3).

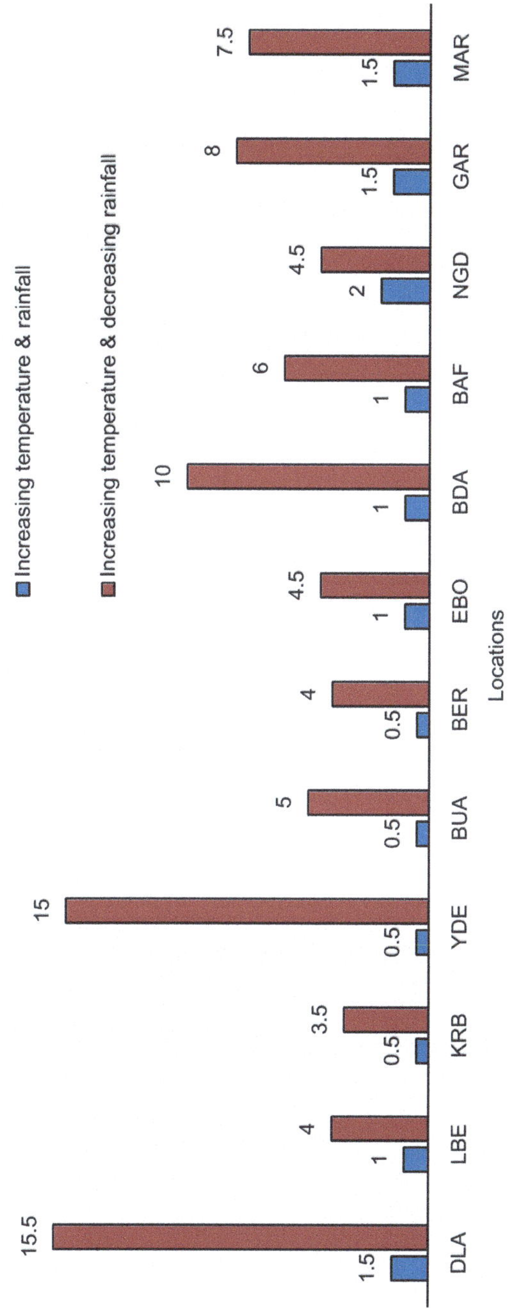

Fig. 2 Climatic hazards in Cameroon

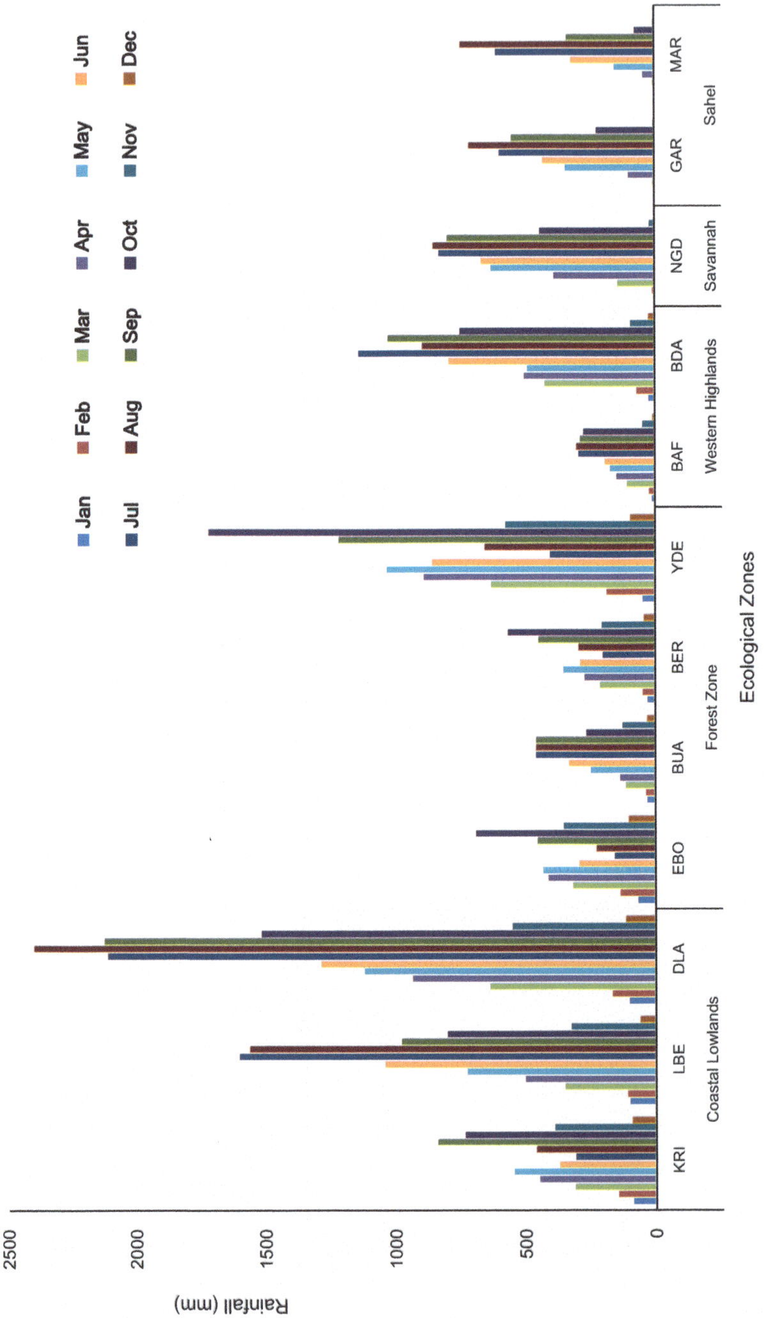

Fig. 3 Mean monthly rainfall patterns in different ecological zones of Cameroon

Kribi, Ebolowa, Bertoua and Yaounde have double rainfall maxima in May/June and September/October. This is typical of the Cameroonian tropical rainfall regime where these localities have two rainy seasons and two dry seasons. The highest amount of rainfall is received in Douala, followed by Limbe. Rainfall reduces gradually northwards, through Buea, Bamenda, Bafoussam, Ngaoundere, Garoua and Maroua. The rainfall characteristics and trends are variable across the country (Table 2).

Meteorological records show variations in rainfall trends over Cameroon. Rainfall trends are relatively stable in Douala, Limbe, Buea, Bamenda and Bafossam (Fig. 4). There have been decreases in Kribi, Yaounde, Bertoua and Ebolowa. Interestingly, the rainfall trend has been increasing in Ngaoundere, Garoua, Maroua. These disparities further elucidate variations in the average amounts of rainfall received over different ecological zones. The coastal lowlands with the proximity to the Atlantic Ocean and the direct influence of monsoonal winds receive an average of 718.92 mm annually (Fig. 5).

The equatorial zones have an average of 362.63 (mm), Western Highlands (334.67 mm), Sudan Savannah (395.97 mm) and Sahel (217.34 mm). Despite these disparities, annual rainfall is still reliable in Cameroon. This was assessed using the Rainfall Coefficient of Variation (RCV) (Fig. 6).

In the tropics, rainfall is unreliable when RCV goes beyond a threshold of 20%. In the ecological zones of Cameroon, the RCV ranges from 6.43% (Western Highlands) to 13.21% (Sahel). This reliability indicates that rainfall in Cameroon is high enough to cause flooding and associated environmental issues in both urban and rural areas. Some areas still receive scanty rain, but when it comes in extreme forms, it results in flooding. A characteristic of a changing climate is the recurrence of extreme weather events [21]. In this context, it is flooding and its associated impacts on urban housing in selected cities in Cameroon (Fig. 7).

Table 2 Rainfall characteristics in Cameroon

Ecological zone	Selected cities	Rainfall characteristics	Observed trends
Coastal Lowlands	DLA	Cameroon Maritime	Stable
	LBE	Cameroon Maritime	Stable
	KRB	Cameroon Maritime	Decrease
Forest Zone	YDE	Equatorial	Decrease
	BUA	Cameroon Maritime	Stable
	BER	Cameroon Maritime	Decrease
	EBO	Cameroon Maritime	Decrease
Western Highlands	BDA	Cameroon Mountain	Stable
	BAF	Cameroon Mountain	Stable
Sudan Savannah	NGD	Humid Tropical	Increase
Sahel	GAR	Dry Tropical (Sahel)	Increase
	MAR	Dry Tropical (Sahel)	Increase

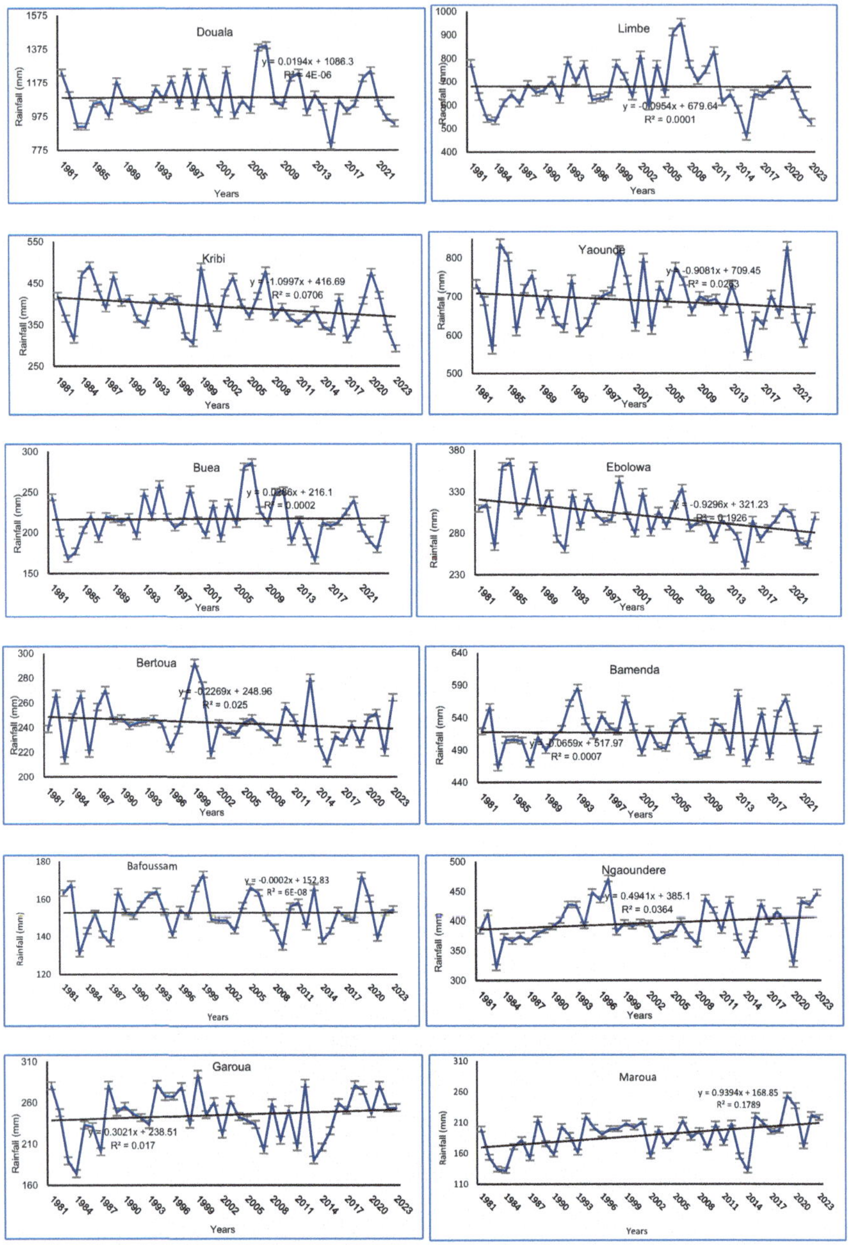

Fig. 4 Rainfall trends across ecological zones of Cameroon (1981–2023)

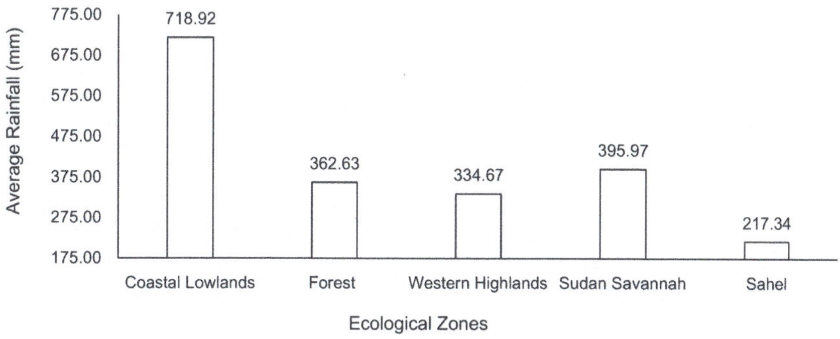

Fig. 5 Average annual rainfall in different ecological zones of Cameroon

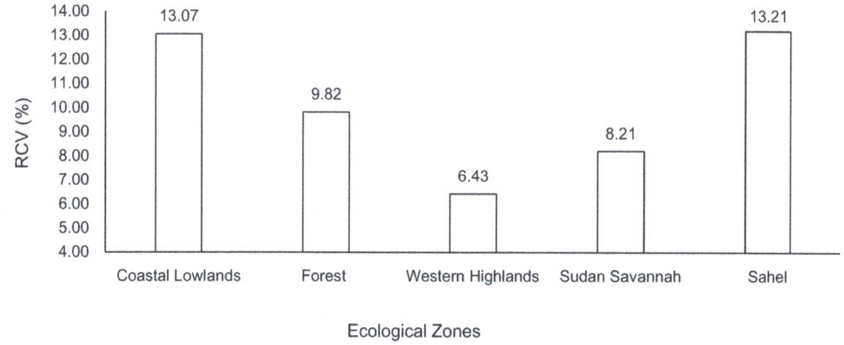

Fig. 6 Rainfall coefficient of variation in Cameroon

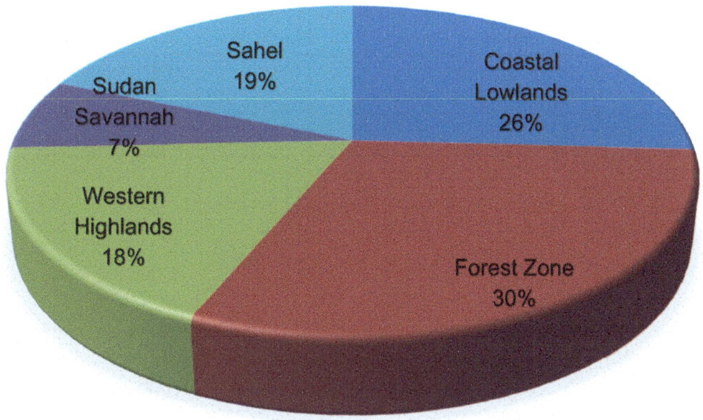

Fig. 7 Changes in intensity of extreme weather events in Cameroon

Results revealed that these extremes are distributed as follows: Coastal Lowlands (26%), Forest Zone (30%), Sahel (19%), Western Highlands (18%) and Sudan Savannah (7%). These extreme weather events have negative repercussions for urban housing development. The drivers of climate change in Cameroon like elsewhere in the tropics are more anthropogenic. Responses in this study indicated that the drivers are deforestation, agriculture, industrialisation and population growth (Table 3).

Deforestation is a major concern in Limbe (substitution of forest cover with palm plantations and settlement), around the suburbs of Yaounde, Ebolawa, Bertoua and in Ngoaundere, Garoua, Maroua. Agriculture is also a leading cause of vegetation loss and this was reported in all the ecological zones, except Ngaoundere. In summary, the relative contribution of climate change drivers is deforestation (32.5%), agriculture (18.5%), industrialisation (10.5%) and population growth (38.5%). The results also indicated that industrial emissions from the oil refineries in Limbe and Kribi, as well as other manufacturing plants in Douala and Bafoussam. Urban population growth is increasing all over the territory, with over 60% of the population already living in urban and semi-urban areas (Fig. 8). The population growth rate in Cameroon was at its peak in the 1990s (3.12%) and it has been projected to reduce to 1950 (1.83%) (Fig. 9).

Population increase is one of the main triggers of waste generation and poor management (a source of local Greenhouse Gas emissions). A significant proportion of this urban population also settles in sub-standard housing that is vulnerable to flooding and other environmental issues. Real-time images of study sites show dense human settlements, with direct implications on the environment (forest loss) (Fig. 10).

Table 3 Drivers of climate change in Cameroon

Town	Deforestation		Agriculture		Industrialisation		Population growth	
	F	%	F	%	F	%	F	%
DLA	10	5	0	0	11	5.5	13	6.5
LBE	5	2.50	2	1.00	1	0.50	2	1.00
KRB	0	0.00	0	0.00	2	1.00	6	3.00
YDE	15	7.50	4	2.00	2	1.00	10	5.00
BUA	0	0.00	6	3.00	0	0.00	5	2.50
BER	7	3.50	1	0.50	0	0.00	1	0.50
EBO	3	1.50	3	1.50	0	0.00	5	2.50
BDA	6	3.00	8	4.00	0	0.00	8	4.00
BAF	0	0.00	5	2.50	5	2.50	4	2.00
NGD	8	4.00	0	0.00	0	0.00	5	2.50
GAR	2	1.00	5	2.50	0	0.00	12	6.00
MAR	9	4.50	3	1.50	0	0.00	6	3.00
Total	65	32.5	37	18.5	21	10.5	77	38.5

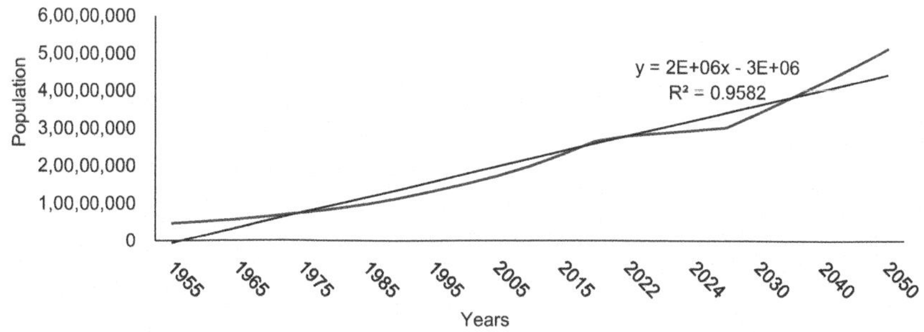

Fig. 8 Population trend in Cameroon (1955–2050). *Data source* https://www.worldometers.info/world-population/cameroon-population/

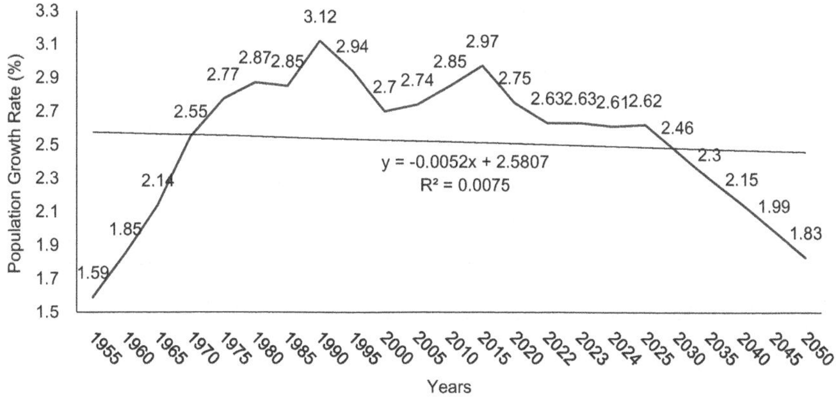

Fig. 9 Population growth rate in Cameroon (1955–2050). *Data source* https://www.worldometers.info/world-population/cameroon-population/

Implications of climate change in Cameroon are extreme weather events, sea level rise, changes in precipitation and temperature patterns, and biodiversity loss. These implications reflect the ecological diversity of the country and each agroecological unit has its uniqueness (Table 4).

Extreme weather events are recurrent in all the ecological zones of the country, as well as changes in precipitation and temperature patterns and biodiversity loss. Extreme weather events like heavy rainfall with floods as the outcomes are common in Douala (3%), Limbe (1.5%), Yaounde (4%), Bamenda (4%), and in the northern regions of the country. Changes in precipitation and temperature patterns have been reported all over the country, especially at the onset of the rainy season [13]. The beginning of the rainy season is often characterised by long dry spells, accompanied

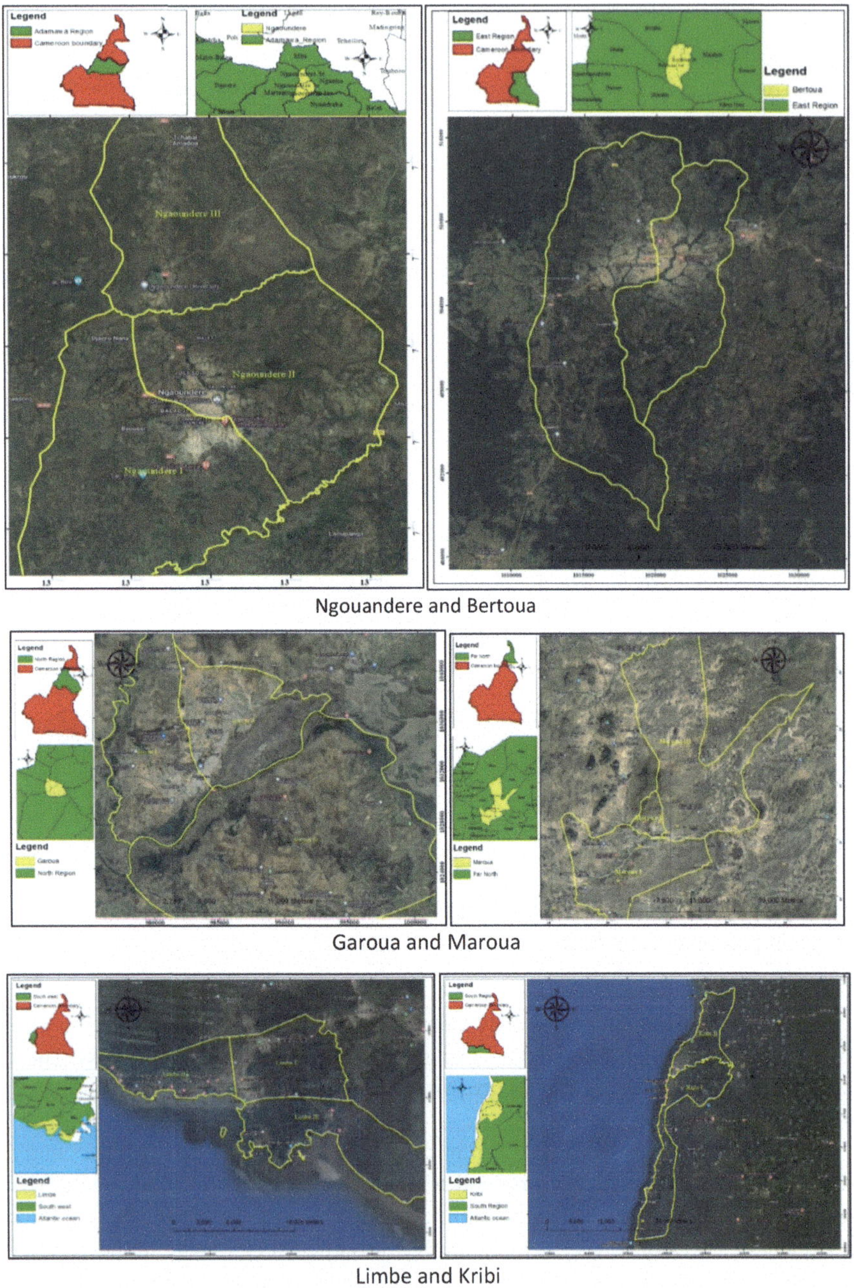

Fig. 10 Real-time images of study sites in Cameroon

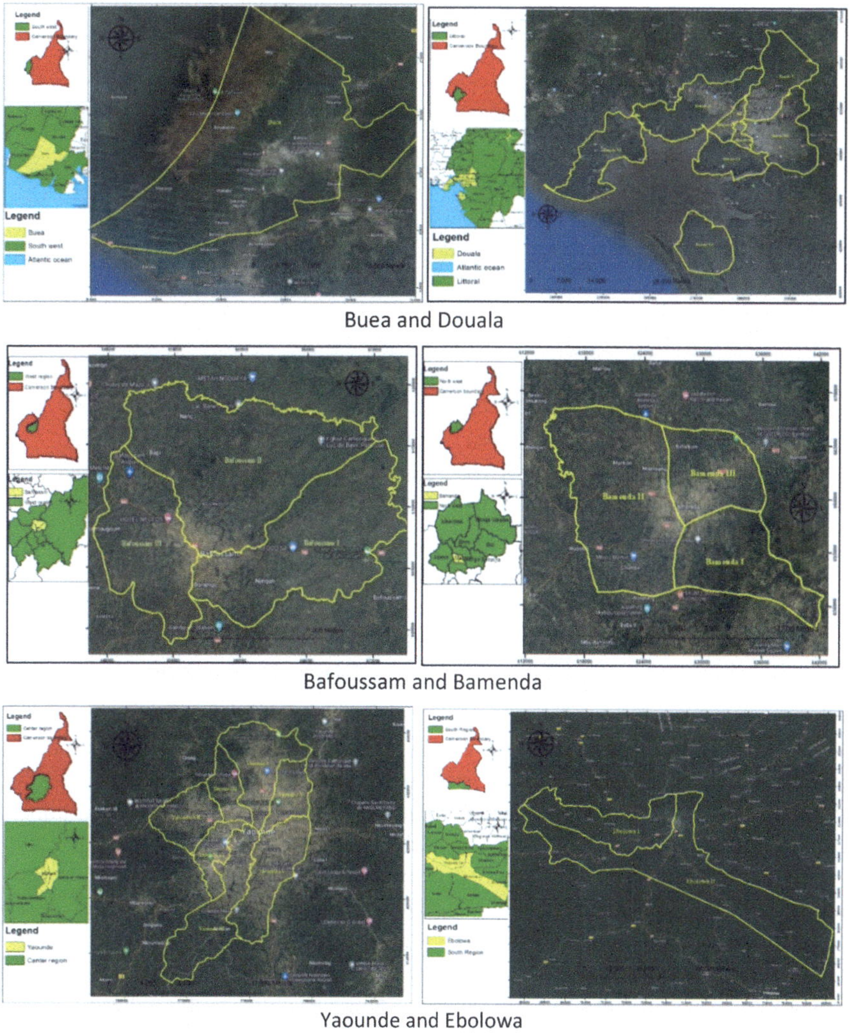

Fig. 10 (continued)

by heat waves and water scarcity. The impacts of sea level rise are felt in Douala, Kribi and Limbe (coastal cities considered in this study) [22]. In all, extreme weather events contribute to 27% of urban vulnerability to climate change, sea level rise (6%), changes in precipitation and temperature patterns (45%) and biodiversity loss (23%).

Table 4 Implications of climate change for urban planning in Cameroon

Cities	Extreme events		Sea level rise		Precipitation & temperature changes		Biodiversity loss	
	F	%	F	%	F	%	F	%
DLA	6	3.00	5	2.50	13	6.50	10	5.00
LBE	3	1.50	2	1.00	4	2.00	1	0.50
KRB	1	0.50	4	2.00	1	0.50	2	1.00
YDE	8	4.00	0	0.00	14	7.00	9	4.50
BUA	2	1.00	0	0.00	6	3.00	3	1.50
BER	0	0.00	0	0.00	5	2.50	4	2.00
EBO	4	2.00	0	0.00	6	3.00	1	0.50
BDA	8	4.00	0	0.00	8	4.00	6	3.00
BAF	3	1.50	0	0.00	10	5.00	1	0.50
NGD	4	2.00	0	0.00	6	3.00	3	1.50
GAR	5	2.50	0	0.00	12	6.00	2	1.00
MAR	10	5.00	0	0.00	5	2.50	3	1.50
Total	54	27	11	6	90	45	45	23

3.2 *Socio-Economic Vulnerability and Urban Housing*

Socio-economic-driven and environmental vulnerability of urban areas in Cameroon is elucidated through vulnerable infrastructure, population vulnerability, vulnerable urban ecosystems, and unsustainable development practices. All these vulnerabilities have been reported in all the study sites (Table 5).

The most vulnerable urban infrastructures in Cameroon are roads (severely degraded in Douala, Yaounde, Bamenda, and Garoua), as well as ailing water infrastructure that can no longer support the increasing population. Striving urban ecosystems such as streams and remnants of vegetation are under severe pressure from the population. Urban streams have become dumping grounds for urban plastic wastes and industrial effluents, making them unfit for use for any form of economic activity, as well as endangering aquatic life. All these are indicators of unsustainable urban development practices, which makes urban areas in Cameroon to be more of '*brown surfaces*', rather than potential green spaces. All of these lead to the vulnerability of the urban population, who are reaping the fruits of their uncivil action in urban areas. In summary, vulnerable infrastructure stands at 27%, vulnerable population (26%), vulnerable ecosystems (35%) and unsustainable development practices (1%). Information gaps in climate action in urban areas in Cameroon include inadequate data on climate change trends and projections, missing links on the vulnerability of ecological zones and communities, and inadequate research on the effectiveness of adaptation and mitigation measures. These gaps have been reported in all the ecological zones of Cameroon (Table 6).

Table 5 Consequences of socio-economic driven climate change vulnerability in Cameroon

Cities	Vulnerable infrastructure		Vulnerable population		Vulnerable ecosystems		Unsustainable development practices	
	F	%	F	%	F	%	F	%
DLA	12	6.00	6	3.00	13	6.50	3	0.15
LBE	3	1.50	3	1.50	3	1.50	1	0.05
KRB	2	1.00	1	0.50	4	2.00	1	0.05
YDE	13	6.50	4	2.00	12	6.00	2	0.1
BUA	2	1.00	2	1.00	5	2.50	2	0.1
BER	3	1.50	2	1.00	3	1.50	1	0.05
EBO	3	1.50	2	1.00	4	2.00	2	0.1
BDA	4	2.00	8	4.00	6	3.00	4	0.2
BAF	0	0.00	8	4.00	4	2.00	2	0.1
NGD	3	1.50	2	1.00	4	2.00	4	0.2
GAR	5	2.50	7	3.50	4	2.00	3	0.15
MAR	3	1.50	6	3.00	8	4.00	1	0.05
Total	53	27	51	26	70	35	26	1

Table 6 Information gaps to climate action in urban areas in Cameroon

Cities	Inadequate data on climate change trends & projections		Missing links on vulnerability of ecological zones & communities		Inadequate research on the effectiveness of adaptation & mitigation measures	
	F	%	F	%	F	%
DLA	13	6.50	10	5.00	11	5.50
LBE	6	3.00	2	1.00	2	1.00
KRB	3	1.50	2	1.00	3	1.50
YDE	16	8.00	5	2.50	10	5.00
BUA	4	2.00	1	0.50	6	3.00
BER	4	2.00	1	0.50	4	2.00
EBO	4	2.00	3	1.50	4	2.00
BDA	8	4.00	8	4.00	6	3.00
BAF	2	1.00	2	1.00	10	5.00
NGD	3	1.50	9	4.50	1	0.50
GAR	3	1.50	3	1.50	13	6.50
MAR	5	2.50	7	3.50	6	3.00
Total	71	35.50	53	26.50	76	38.00

Inadequate data on climate change trends and projections is a serious research gap in Cameroon. A study of this magnitude needed data on other climatic elements like temperature and relative humidity (which are practically unavailable in the National Meteorological Service), as well as in-depth data on flood incidents and impacts. There are also missing links on the vulnerability of ecological zones and communities, as well as inadequate research on the effectiveness of adaptation and mitigation measures. This study will likely fill some of these gaps and pave the way for future pathways. In all inadequate data on climate change trends and projections stands at 35.5%, missing links on the vulnerability of ecological zones and communities (26.5%) and inadequate research on the effectiveness of adaptation and mitigation measures (38%). The data gaps in climate action have necessitated the use of indigenous knowledge systems to adapt to changing climatic conditions in urban and rural areas of Cameroon [23]. This dependence on traditional ecological knowledge depends on insights from past adaptations, culturally appropriate and sustainable strategies, and building resilience at the community level (Table 7).

Cameroon is a culturally rich and diverse country where culturally appropriate and sustainable strategies are employed in the management of resources like land, forests, and water all over the national territory. Adaptation practices are linked to culture and have been handed down from one generation to another, using indigenous knowledge systems. This has enhanced the building of resilience to climate change at the community level. Insights from past adaptations stand at 13%, culturally appropriate and sustainable strategies (56%), and building resilience at the community level (31%). Faced with increasing urban climate vulnerability and inadequate data to address the emerging problems, local communities have been engaged in data

Table 7 Traditional knowledge in climate adaptation in urban Cameroon

Cities	Insights from past adaptations		Culturally appropriate & sustainable strategies		Building resilience at the community level	
	F	%	F	%	F	%
DLA	2	1.00	24	12.00	8	4.00
LBE	1	0.50	4	2.00	5	2.50
KRB	3	1.50	4	2.00	1	0.50
YDE	2	1.00	21	10.50	8	4.00
BUA	0	0.00	7	3.50	4	2.00
BER	0	0.00	8	4.00	1	0.50
EBO	2	1.00	7	3.50	2	1.00
BDA	6	3.00	8	4.00	8	4.00
BAF	0	0.00	12	6.00	2	1.00
NGD	4	2.00	3	1.50	6	3.00
GAR	1	0.50	6	3.00	12	6.00
MAR	5	2.50	8	4.00	5	2.50
Total	26	13.00	112	56.00	62	31.00

collection on climate change impacts, reporting impacts to relevant stakeholders and participating in adaptation and mitigation measures (Table 8). Community participation in data collection on the impacts of climate change stands at 8.5%, reporting impacts to relevant stakeholders (20%) and participation in adaptation and mitigation measures (71.5%).

Communication is vital to translating climate action into a development agenda. For this to happen, climate change communication should be in clear and concise language, use visuals, tailoring climate action to stakeholder needs and interest, as well use of different communication channels (Table 9).

Even residents in rural areas have been quick to note that abnormal weather behaviour is a sign of a changing climate. There are signs and indicators of these changes in all the ecological zones of Cameroon. What differs in the way in which the local changes are communicated? For climate change communication to be meaningful, it must use clear and concise local languages (13.5%), use visuals to illustrate local realities (30.5%), tailor communication to stakeholder needs and interests (21%) and above all use various communication channels (traditional and modern media) (35%), like community and national radio stations, local television channels, social media handles, local newspaper columns, social gatherings like churches and mosques, open market places, public transportation systems, schools (pre-nursery to universities), as well as all other channels where people can easily be informed on the devastating consequences of climate change.

The changing climate results in recurring heat waves and flooding in temperate regions, diminishing water tables and extreme weather in sub-tropical savannahs, and water scarcity in urban centres of Sub-Saharan Africa. Coastal towns are also

Table 8 Role of local communities in climate action in urban Cameroon

Cities	Data collection on climate change impacts		Reporting impacts to relevant stakeholders		Participation in adaptation and mitigation measures	
	F	%	F	%	F	%
DLA	2	1.00	4	2.00	28	14.00
LBE	1	0.50	4	2.00	5	2.50
KRB	1	0.50	2	1.00	5	2.50
YDE	2	1.00	4	2.00	25	12.50
BUA	0	0.00	1	0.50	10	5.00
BER	0	0.00	0	0.00	9	4.50
EBO	2	1.00	1	0.50	8	4.00
BDA	6	3.00	8	4.00	8	4.00
BAF	0	0.00	3	1.50	11	5.50
NGD	0	0.00	3	1.50	10	5.00
GAR	1	0.50	3	1.50	15	7.50
MAR	2	1.00	7	3.50	9	4.50
Total	17	8.50	40	20.00	143	71.50

Table 9 Best practice for climate change communication in Cameroon

Cities	Clear and concise language		Use of visuals		Tailoring to stakeholder needs &interests		Use of various communication channels	
	F	%	F	%	F	%	F	%
DLA	2	1.00	16	8.00	4	2.00	12	6.00
LBE	2	1.00	2	1.00	1	0.50	5	2.50
KRB	1	0.50	2	1.00	3	1.50	2	1.00
YDE	3	1.50	13	6.50	4	2.00	11	5.50
BUA	0	0.00	4	2.00	4	2.00	3	1.50
BER	0	0.00	4	2.00	1	0.50	4	2.00
EBO	3	1.50	3	1.50	3	1.50	2	1.00
BDA	8	4.00	4	2.00	6	3.00	4	2.00
BAF	1	0.50	5	2.50	3	1.50	5	2.50
NGD	3	1.50	3	1.50	0	0.00	7	3.50
GAR	2	1.00	3	1.50	8	4.00	6	3.00
MAR	2	1.00	2	1.00	5	2.50	9	4.50
Total	27	13.50	61	30.50	42	21.00	70	35.00

contending with increased flooding due to these climatic shifts. These changes disproportionately affect impoverished urban residents in slums and underdeveloped areas, exacerbating the strain on settlements and prolonging recovery periods. In addition to addressing urban challenges such as poverty, unemployment, poorly planned construction, and decreasing urban greenery, developing economies are grappling with the impact of climate change.

Residential housing, a dominant fabric of urban centres occupying geographical risk zones, is an object of potential risk from climatic hazards such as landslides, erosion, floods, and extreme heat. To comprehend the types of climatic hazards in the urban settlements of Cameroon, the major ecological zones; the Coastal Lowlands, the Forest Zone, the Western Highlands, Sudan Savannah, and Sahel were considered. The major climatic hazards in the urban residential housing of Cameroon include flooding, landslides, and extreme sea level rise (Fig. 11).

A major challenge to reckon with is the degrading effects of both hydrological and climatological hazards, in which flooding and extreme heat occurred at high magnitude in the Coastal Lowlands and Forest Zone and Sahelian Region respectively. Major coastal towns of the country such as Douala and Limbe bordered by the Atlantic Ocean and have poor drainage systems risk flooding in the event of a heavy downpour. For example, on the morning of 21st August 2020, the neighbourhoods of Makepe Missoke, Bonaberi, and Bonamoussadi were flooded with many homes submerged, likewise the occurrence of flooding and landslide in Lime that left five deaths and approximately 2000 people affected. The rising sea level poses a great threat to settlements in the coastal towns of Cameroon such as Douala, Kribi, and

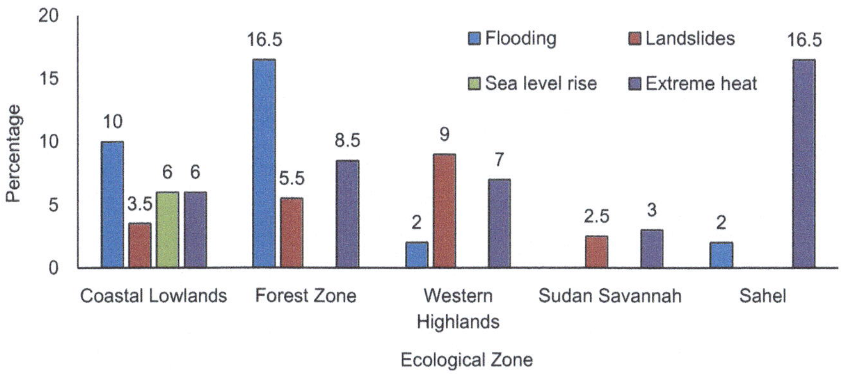

Fig. 11 Climatic hazards in urban residential housing in Cameroon

Lime given that the management plan for coastal areas of Cameroon has not been fully designed and implemented.

The gradual occupation of the hillslopes in the Western Highlands which is situated along the Cameroon Geologic Line of Weakness puts settlements at risk of landslides and erosion. In this manner, heavy rainfall on 29th of October, 2019 caused a landslide that killed 42 persons and left hundreds in the Ngouache Neighbourhood of Bafoussam. Flooding occurs not only in the coastal zones but in inland towns with poor natural drainage and poor urban planning such as Bamenda in the Western Highlands and Yaounde located far inland. Equally, of major concern is the extreme heat witnessed by residents in all the ecological zones where the Sahel, Semi-arid, experienced the highest temperatures. Apart from the discomfort caused by extreme heat in the Sahel Region, the short-rain season is always torrential, causing the overflow of the Chari and Logone. For example, the torrential rainfall caused the collapse of the bridge affecting the Logone and Chari, Mayo Danay, Mayo Kani Division, and the torrential rainfall of 11th and 12th of September 2020 which affected more than 26,402 people and 3152 households (International Federation of Red Cross and Red Crescent, 2021).

More communities are likely going to be affected by climate change hazards depending on their resilience, economic strength, level of technology, availability of preplanning, distance from the hazard location of the community, and access to resources. The socio-economic vulnerability indicators in the different ecological zones of Cameroon included low-income households, limited resources for retrofitting homes, and limited access to basic services (Fig. 12). Urban communities are more vulnerable to the impacts of climate change largely due to low-income households who are the occupants of disaster-prone zones and often have limited access to resources such as finances, materials, and expertise for housing revitalization [6]. The low-income households live in sub-standard housing lacking services that can help regulate high temperatures such as alternating current systems and planned home gardens.

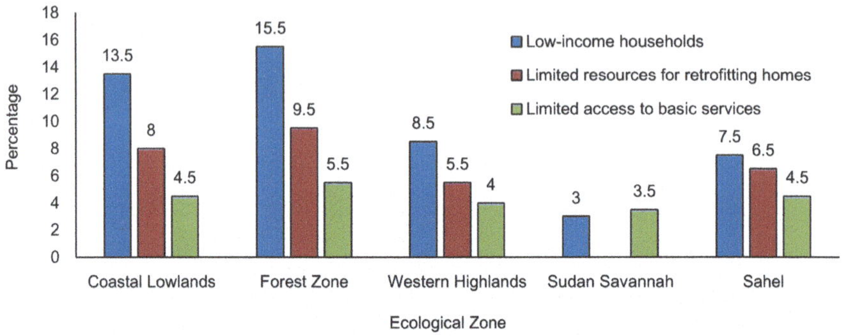

Fig. 12 Socio-economic vulnerability of urban housing to climate change in Cameroon

The dilemma is the fact the hard hit by climate change impacts are the poor and deprived communities of the urban centres whose resilience and safety are at the mercy of flooding, landside, extreme heat, and pollution. The three indicators imply the urgency for institutional, physical, socio-economic, and culturally inclined mechanisms to mitigate the impacts of climate variability in the lagging and vulnerable communities, particularly in Sub-Saharan Africa [24, 25].

The rapid urbanisation of the unplanned urban centres is the major cankerworm that eats up settlements on marginal lands. The damaging effects of climate change exacerbate the suffering of households living especially in the geographically disadvantageous areas of the cities. Understanding climate change for mitigation and adaptation requires full knowledge of the baseline situation, the state of the community, their strength, weaknesses, and opportunities provided by the physical and the human environment [26]. Climate change poses great threads of stress and depression in less resilient communities which is of great concern to town planning and urban development. The consequences of climate change on urban residential housing in Cameroon included insights from coping with climate variability, identification of vulnerable areas and communities, and development and implementation of culturally appropriate and sustainable adaptation strategies (Table 10).

The result showed poor insights from coping with climate change (35%), where Douala recorded a higher percentage of 8% and Bafoussam and Ebolowa with the least of 0.50%, poor (25%) mapping of vulnerable areas and communities, and poor development and implementation of desired cultural and sustainable adaptation strategies (39%). Though with variation in the three indicators for different towns, little knowledge exists about how communities and individual households stress and respond to such stress, the perception of communities towards climate change impacts on their living environment, and the location of vulnerable communities and their economic, social, and demographic characteristics. The poor identification of vulnerable areas and communities only permits the inappropriate development and implementation of local and nature-based approaches for mitigation and adaptation [26]. The baseline is the identification of climate change risk settlements, listening

Table 10 Consequences of climate change for urban residential housing in Cameroon

Cities	Insights from coping with climate variability		Identification of vulnerable areas & communities		Development & implementation of culturally appropriate & sustainable adaptation strategies	
	F	%	F	%	F	%
DLA	16	8.00	6	3.00	12	6.00
LBE	4	2.00	2	1.00	4	2.00
KRB	2	1.00	2	1.00	4	2.00
YDE	13	6.50	6	3.00	12	6.00
BUA	4	2.00	4	2.00	3	1.50
BER	4	2.00	1	0.50	4	2.00
EBO	1	0.50	4	2.00	6	3.00
BDA	8	4.00	8	4.00	6	3.00
BAF	1	0.50	2	1.00	11	5.50
NGD	7	3.50	3	1.50	3	1.50
GAR	7	3.50	5	2.50	7	3.50
MAR	3	1.50	8	4.00	7	3.50
Total	70	35.00	51	25.50	79	39.50

to their stories of stresses, how they cope, and documenting them. However, this is lacking in urban Cameroon given the little efforts made in collecting such data.

Climate change impacts are very devastating so much so that the victim communities often exhaust their capacities and would require external support. In some cases, the impacts may be slowly occurring such as the drying up of streams, and fluctuation in annual rainfall amount discouraging the zeal for urgent intervention, while in some it may be the urban dwellers in case of rapidly occurring hazards such as flooding of housing, landslide, and extreme heat pushing the communities beyond the threshold. Local institutions are likely in a better position to understand to provide information on vulnerability. The role of local institutions in vulnerability mapping was categorized into data collection on vulnerable areas, participation in developing vulnerability maps, and the use of maps in adaptation, planning, and decision-making (Table 11). The local institutions such as the traditional institutions, community associations, and locally based NGOs are close to the communities but have played little role in vulnerability mapping (0.50%) for the city of Bertoua, and the highest (7%) for the city of Douala.

The participation of local institutions in the development of vulnerable areas and the use of maps in adaptation planning and decision-making is shallow but significant in reducing vulnerabilities in residential housing in Cameroon, as postulated by Girma et al. [27]. The overall engagement of local institutions in vulnerability mapping (34%), participation in the development of vulnerability maps (29.5%), and the use of maps in the adaptation and decision-making process (36%) implies

Table 11 Role of local institutions in mapping vulnerability in Cameroon

Cities	Data collection on vulnerable areas		Participation in the development of vulnerability maps		Use of maps in adaptation planning & decision-making	
	F	%	F	%	F	%
DLA	14	7.00	6	3.00	14	7.00
LBE	4	2.00	3	1.50	3	1.50
KRB	2	1.00	1	0.50	5	2.50
YDE	11	5.50	7	3.50	13	6.50
BUA	2	1.00	4	2.00	5	2.50
BER	3	1.50	1	0.50	5	2.50
EBO	1	0.50	5	2.50	5	2.50
BDA	8	4.00	8	4.00	6	3.00
BAF	9	4.50	3	1.50	2	1.00
NGD	5	2.50	4	2.00	4	2.00
GAR	5	2.50	9	4.50	5	2.50
MAR	5	2.50	8	4.00	5	2.50
Total	69	34.50	59	29.50	72	36.00

little participation of communities improving their living housing environment and other related development projects. The sidelining of local institutions regarding the threats posed by climate change to residential housing and communities' resilience limits the accuracy of project implementation at the local levels which can only be solved bottom-top approach and or community participation approach [28].

Geomorphic processes have created variations in physical landscape which in turn influence the location of settlements and the risk associated with those settlements. Urban centres situated in low and coastal areas are prone to the risk of flooding and pollution of water resources, whilst settlements on plateaus and hillslopes are more exposed to the risk of landslides mudflow, and erosion. The impacts of these hazards are exacerbated by climate change which increases the environmental, socio-economic, and cultural vulnerabilities. The dominance of key socio-economic, and climate change vulnerabilities and little participation of the local institution in the overall objective of sustainable adaptation in all the ecological zones of Cameroon calls for an institutional, socio-economic, cultural, and administrative approach to improve the living conditions and standard of living and both the natural and cultural landscape in Sub-Saharan Africa.

3.3 *Adaptation Strategies and Sustainable Architectural Designs*

Adaptation and sustainable architectural designs are critical in mitigating the urban climate crisis, especially in vulnerable regions like Cameroon [29]. By integrating strategies such as raising building foundations, using weather-resistant materials, planting vegetation for shade, and implementing rainwater harvesting and renewable energy systems, cities can enhance their resilience to climate-related challenges. This approach not only addresses immediate vulnerabilities, such as flooding and extreme weather, but also promotes long-term sustainability by aligning urban development with the realities of a changing climate. Some sustainable architectural designs are implemented across major Cameroonian cities. These include raising foundations to reduce flooding, using weather-resistant materials, planting vegetation for shade, installing rainwater harvesting systems, and using renewable energy sources (Table 12). It showcases the varying adoption rates of different strategies, providing insights into local priorities and challenges.

Results reveal that raising foundations to reduce flooding is the most widely adopted sustainable design measure, with 26.5% overall implementation. Douala (7%) and Yaounde (5.5%) lead in this strategy, likely due to their vulnerability to flooding. The use of weather-resistant materials follows closely at 20.5% overall, with Yaounde and Garoua (both 3.5%) showing the highest adoption rates. Planting vegetation for shade is particularly prevalent in Maroua (5.5%) and Yaounde (2.5%),

Table 12 Sustainable architectural designs in Cameroon

Cities	Raising foundations to reduce flooding		Using weather-resistant materials		Planting vegetation for shade		Installing rainwater harvesting systems		Using renewable energy sources	
	F	%	F	%	F	%	F	%	F	%
DLA	14	7.00	6	3.00	6	3.00	4	2.00	4	2.00
LBE	4	2.00	3	1.50	0	0.00	2	1.00	1	0.50
KRB	2	1.00	0	0.00	0	0.00	2	1.00	4	2.00
YDE	11	5.50	7	3.50	5	2.50	4	2.00	4	2.00
BUA	2	1.00	2	1.00	1	0.50	2	1.00	4	2.00
BER	3	1.50	1	0.50	4	2.00	0	0.00	1	0.50
EBO	1	0.50	5	2.50	0	0.00	2	1.00	3	1.50
BDA	4	2.00	4	2.00	4	2.00	6	3.00	4	2.00
BAF	5	2.50	3	1.50	2	1.00	3	1.50	1	0.50
NGD	0	0.00	0	0.00	4	2.00	3	1.50	6	3.00
GAR	5	2.50	7	3.50	1	0.50	3	1.50	3	1.50
MAR	2	1.00	3	1.50	11	5.50	0	0.00	2	1.00
Total	53	26.50	41	20.50	38	19.00	31	15.50	37	18.50

reflecting local climate conditions. Rainwater harvesting systems and renewable energy sources are less commonly implemented, at 15.5% and 18.5% respectively, indicating potential areas for future development. The perceived effectiveness of specific design interventions in reducing vulnerability to climate impacts in the urban housing sector captured in this study include raising foundations against floods, weather-resistant materials, culverts, and drains (Table 13).

This assessment highlights which measures are considered most valuable in different urban contexts. Raising foundations against floods and using weather-resistant materials are equally valued by respondents (both at 41.5% overall), with Douala and Yaounde residents showing the highest appreciation for these measures. This suggests that direct structural interventions are perceived as highly effective in mitigating climate impacts. Culverts and drains, while seen as less effective overall (17%), are more valued in cities like Bamenda and Garoua (both 3%), possibly due to specific local water management needs. Prioritizing climate adaptation benefits across Cameroon's ecological zones was captured in resilience, improvement of quality of life, and job security. It provides a visual representation of how different regions value various aspects of climate resilience and development (Fig. 13).

The prioritisation of climate adaptation benefits varies significantly across ecological zones. The Forest Zone and Coastal Lowlands heavily emphasize resilience to climate change (75% and 70% respectively), reflecting their high vulnerability to environmental shifts and risks like sea-level rise. The Western Highlands show a balanced approach, with equal emphasis (45%) on improving quality of life and climate resilience, indicating a holistic view of adaptation. In contrast, the Sudan Savannah and Sahel zones prioritize job security (30%) and quality of life (40%),

Table 13 Effectiveness of designs in reducing urban housing vulnerability

Cities	Raising foundations against floods		Weather-resistant materials		Culverts & drains	
	F	%	F	%	F	%
DLA	20	10.00	12	6.00	2	1.00
LBE	4	2.00	3	1.50	3	1.50
KRB	3	1.50	3	1.50	2	1.00
YDE	17	8.50	12	6.00	2	1.00
BUA	4	2.00	4	2.00	3	1.50
BER	3	1.50	5	2.50	1	0.50
EBO	5	2.50	5	2.50	1	0.50
BDA	6	3.00	10	5.00	6	3.00
BAF	5	2.50	6	3.00	3	1.50
NGD	6	3.00	3	1.50	4	2.00
GAR	4	2.00	9	4.50	6	3.00
MAR	6	3.00	11	5.50	1	0.50
Total	83	41.50	83	41.50	34	17.00

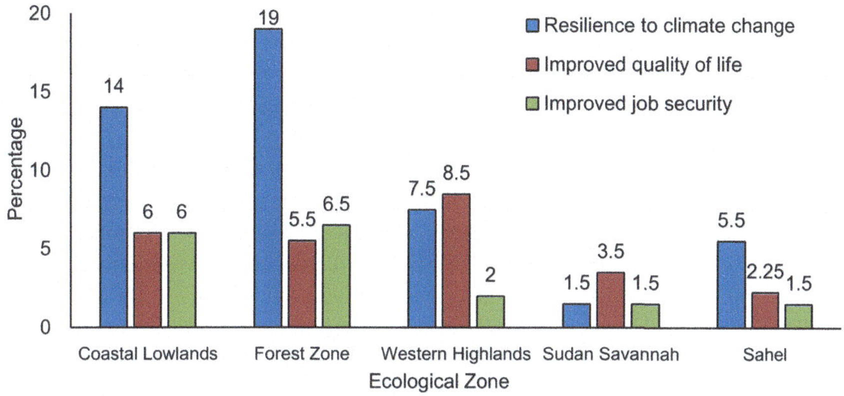

Fig. 13 Opportunities for climate action in Cameroon

suggesting that immediate socio-economic concerns take precedence over long-term climate resilience in these regions.

Stakeholder roles in climate adaptation for sustainable housing examine the multifaceted approach required to address climate change impacts on urban housing in Cameroon. It recognizes that effective adaptation strategies necessitate the involvement of various stakeholders (state, private sector, civil society organisations, and local communities), each playing a crucial role in the process. By exploring these diverse roles, the analysis provides a clearer understanding of the collaborative climate adaptation in the context of sustainable housing, emphasizing the importance of coordinated efforts among different actors to achieve resilience and sustainability in urban environments [30]). The role of the state in climate adaptation shows variations in government involvement across different ecological zones. The measurable indicators captured in this study are policies and regulations, financial and technical assistance, and raising awareness (Fig. 14).

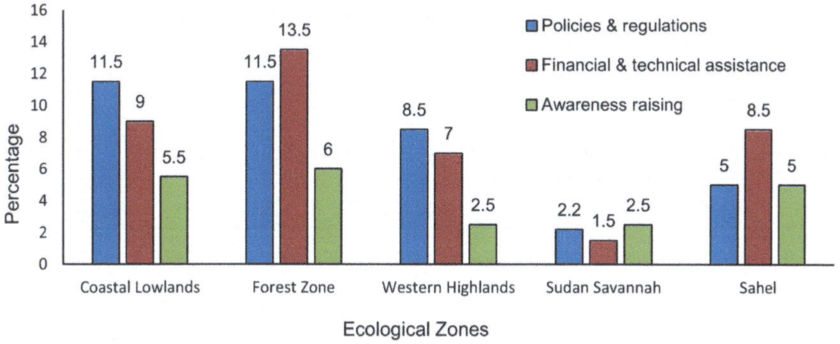

Fig. 14 Role of the state in climate adaptation in the housing sector in Cameroon

State involvement in climate adaptation varies across zones, with the highest engagement in the Forest Zone and Western Highlands. These areas see significant state efforts in policies and regulations (11.5% and 8.5% respectively) and financial and technical assistance (13.5% and 7%). The Coastal Lowlands also show strong state involvement, particularly in policies (11.5%). The Sudan Savannah and the Sahel have lower state engagement overall, indicating a potential need for increased government attention in these regions to support climate adaptation efforts. Private sector initiatives focus on demonstrating how business involvement in climate adaptation varies by region. It offers insights into potential areas for increased private sector participation through developing market-supportive products, investing in research and development and partnerships between the government and civil society (Fig. 15).

The private sector's role is most pronounced in the Forest Zone and Coastal Lowlands, with a strong focus on developing and marketing supportive products (20% and 18.5% respectively). These areas also see higher investment in research and development. The Western Highlands show moderate private sector engagement across all activities. The Sudan Savannah and Sahel have lower private sector involvement, suggesting an opportunity for increased business participation in climate adaptation strategies in these regions. Civil society's contribution to climate adaptation efforts is showcased in raising awareness, capacity building and training, as well as advocacy (Fig. 16). This visualisation helps identify where non-governmental organisations are most active and where their presence could be strengthened.

Civil society organisations play a crucial role across all zones, with the Western Highlands showing the highest overall engagement. Capacity building and training are the primary focus in most zones, particularly in the Western Highlands (28.5%) and Forest Zone (16.5%). Awareness raising is also a significant activity, especially in the Western Highlands (20.5%). The Sudan Savannah shows the lowest civil society engagement, indicating a potential gap in grassroots climate adaptation efforts in this region. As much as the state, private sector and civil society are active in climate action, community responses to climate adaptation needs should be reasonable and

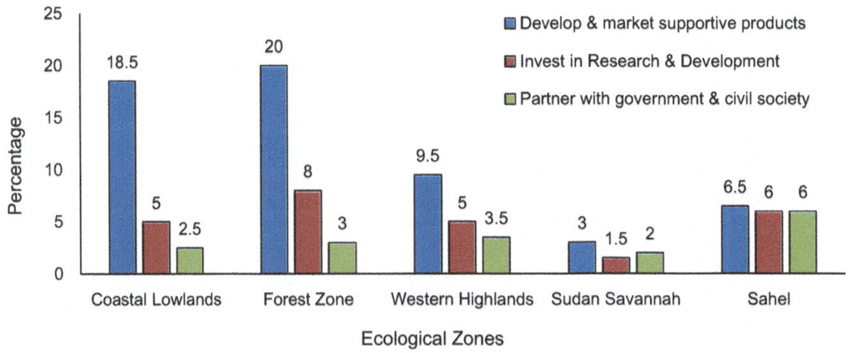

Fig. 15 Private sector initiatives for climate adaptation in the housing sector in Cameroon

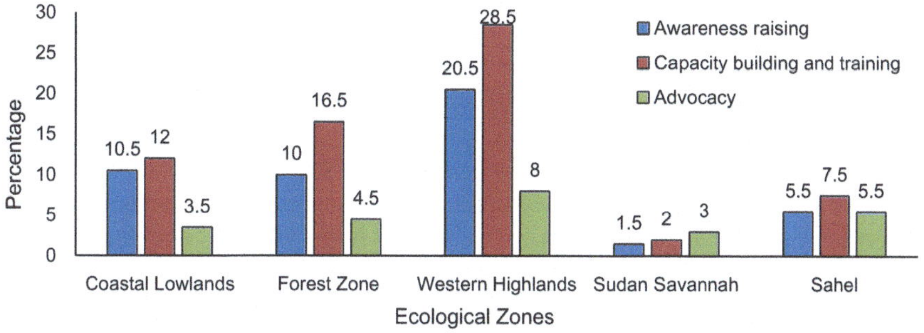

Fig. 16 The role of civil society in climate adaptation in the housing sector Cameroon

proactive. Some of these are the demand for sustainable housing designs, shared knowledge and active participation in development (Fig. 17).

Community engagement in climate adaptation varies across zones. The Forest Zone shows the highest overall community involvement, particularly in demanding sustainable designs (13%). The Coastal Lowlands and the Sahel also demonstrate significant community participation. The Western Highlands and Sudan Savannah show lower levels of community engagement, suggesting a need for increased community mobilisation and empowerment in these areas. This comprehensive analysis reveals that climate adaptation strategies and stakeholder involvement vary significantly across Cameroon's ecological zones. The Forest Zone and Coastal Lowlands generally show higher engagement across all stakeholder groups, likely due to their increased vulnerability to climate impacts.

The Western Highlands demonstrate a balanced approach, while the Sudan Savannah and Sahel regions consistently show lower engagement levels, indicating a need for increased focus and resources in these areas. These disparities highlight the importance of tailoring climate adaptation and sustainable housing strategies to the

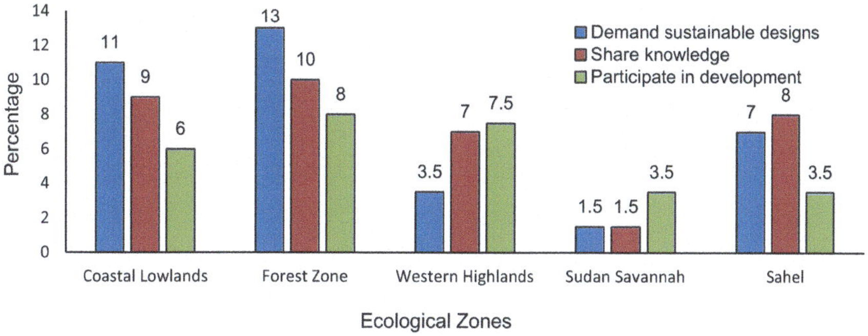

Fig. 17 Community responses to climate action in the housing sector in Cameroon

specific ecological, social, and economic contexts of each region. These adaptation strategies will be more robust if Nature-based Solutions and Ecosystem-based Adaptation are integral components of climate solutions. Nature-based Solutions (NbS) and Ecosystem-based Adaptation (Eba) can contribute over 30% of the total climate change mitigation required by 2030 to retain warming below 2 °C [19]. Inspired by natural environmental practices and processes, NbS contributes to ecosystem management. This involves the rehabilitation of ecosystems and regenerating natural processes in man-made ecosystems [19]. EbA is an important sub-component of NbS which helps people adapt to climate change [19]. Such nature-based approaches use ecosystem services to adapt to the hostile effects of climate change. Some EbA solutions include landscape restoration. EbA relies on local ecological knowledge to address the local challenges of climate change [19].

3.4 Policy Implications and Stakeholder Engagement

Policy priorities for sustainable urban housing in Cameroon captured in this study include a national policy on climate change adaptation, financial & technical assistance, integration into urban development policies and research and development of new technologies outlined in (Table 14). This information is crucial for understanding the direction of future urban development and climate adaptation efforts.

Table 14 Policy priorities for sustainable housing in Cameroon

Cities	National policy on climate change adaptation		Financial & technical assistance		Awareness raising		Integration into urban development policies		Research & development of new technologies	
	F	%	F	%	F	%	F	%	F	%
DLA	6	3.00	2	1.00	5	2.50	13	6.50	8	4.00
LBE	2	1.00	2	1.00	1	0.50	2	1.00	3	1.50
KRB	2	1.00	1	0.50	1	0.50	3	1.50	1	0.50
YDE	7	3.50	2	1.00	5	2.50	9	4.50	8	4.00
BUA	2	1.00	1	0.50	1	0.50	5	2.50	2	1.00
BER	1	0.50	1	0.50	1	0.50	5	2.50	1	0.50
EBO	6	3.00	0	0.00	1	0.50	2	1.00	2	1.00
BDA	2	1.00	6	3.00	4	2.00	6	3.00	4	2.00
BAF	6	3.00	2	1.00	2	1.00	3	1.50	1	0.50
NGD	2	1.00	0	0.00	5	2.50	3	1.50	3	1.50
GAR	3	1.50	7	3.50	1	0.50	6	3.00	2	1.00
MAR	7	3.50	1	0.50	1	0.50	3	1.50	6	3.00
Total	46	23.00	25	12.50	28	14.00	60	30.00	41	20.50

Policy priorities for sustainable urban housing reveal that the integration of urban development policies is at the top (30% overall), with Douala (6.5%) and Yaounde (4.5%) leading in emphasizing this aspect, reflecting the need for comprehensive urban planning approaches. National policy on climate change adaptation is the second-highest priority (23%), with strong support in Yaounde and Maroua (both 3.5%). Research and development on new technologies is also highly valued (20.5%), particularly in Douala and Yaounde (both 4%), indicating recognition of the role of innovation in addressing climate challenges. Financial and technical assistance (12.5%) and awareness raising (14%) are seen as relatively lower priorities, though still significant in the overall strategy.

4 Discussion and Implications

4.1 Climate Change Trends and Ecological Vulnerability

The study reveals significant climate change trends across Cameroon's ecological zones, with increasing temperatures and varying rainfall patterns being the most prominent. This aligns with Williams et al. [3, 4], who highlighted the mounting global concern of urban vulnerability to climate change, particularly in rapidly urbanizing areas [31]. The findings show that coastal lowlands like Douala and Limbe are experiencing stable rainfall trends, while inland areas like Kribi, Yaounde, Bertoua, and Ebolowa are seeing decreases. Interestingly, northern regions (Ngaoundere, Garoua, and Maroua) are experiencing increasing rainfall trends. These diverse patterns underscore the complex nature of climate change impacts across different ecological zones, as noted by [9] in their study of African cities. The study's revelation of extreme weather events, particularly flooding and its associated impacts on urban housing, corroborates the findings of Yonzoua et al. [8], who identified flooding as a major risk in Cameroonian cities due to rapid and often unplanned urbanisation. The vulnerability is particularly high in slum areas, which make up an estimated 65–70% of urban areas in the country. This highlights the need for targeted interventions in urban planning and development to address these vulnerabilities.

4.2 Socio-Economic Vulnerability and Urban Housing

The study identifies low-income households, limited resources for retrofitting homes, and limited access to basic services as key indicators of socio-economic vulnerability in urban housing across Cameroon. This finding resonates with the work of Ngụ et al. [6], who found that peri-urban and poor households are often the most susceptible to climate change impacts. The study's emphasis on the vulnerability of substandard housing in disaster-prone zones echoes the concerns raised by Tchouongsi

[11] regarding the heightened flood vulnerability in Yaoundé due to demographic pressure and uncontrolled urbanisation. The limited participation of local institutions in vulnerability mapping and adaptation planning, as revealed in the study, points to a significant gap in community engagement. This aligns with the observations of Lueong and Tume [12, 32, 33], who identified ineffective institutional performance and weak legislative frameworks as factors exacerbating flood vulnerability in northern Cameroon. The study's call for a bottom-up approach and community participation in adaptation planning reflects the emerging vulnerability paradigm discussed by [7], which emphasizes the need for comprehensive understanding and targeted adaptation measures.

4.3 Adaptation Strategies and Sustainable Architectural Designs

The study's findings on adaptation strategies, particularly the emphasis on raising foundations to reduce flooding and using weather-resistant materials, demonstrate practical responses to climate risks. This approach aligns with the integrated adaptation strategies suggested by [6], which include investments in infrastructure and poverty reduction initiatives. The varying prioritisation of climate adaptation benefits across ecological zones, from resilience to climate change in coastal areas [31] to job security and quality of life in inland regions, reflects the need for context-specific adaptation strategies as highlighted by [5, 34]. The role of traditional knowledge in adaptation, as identified in the study, adds an important dimension to climate change responses. This aligns with the growing recognition of indigenous knowledge systems in climate adaptation strategies, as noted in broader climate change literature [23]. The emphasis on culturally appropriate and sustainable strategies (56% of responses) underscores the importance of integrating local knowledge and practices into formal adaptation plans.

4.4 Policy Implications and Stakeholder Engagement

The study's findings on policy priorities, particularly the emphasis on integrating climate adaptation into urban development policies (30% of responses), reflect the need for a comprehensive approach to urban planning in the face of climate change [35]. This aligns with the new vulnerability paradigm discussed by [7], which aims to provide city decision-makers with tools to analyze and implement effective adaptation strategies. The varying levels of stakeholder engagement across different ecological zones, as revealed in the study, highlight the need for targeted efforts to enhance participation, especially in the Sudan Savannah and Sahel regions. This finding resonates with the works of Awazi et al. [15] [16], and [11], who emphasized the importance of understanding local contexts and engaging communities in climate adaptation efforts. The study provides a comprehensive overview of climate change impacts on urban housing in Cameroon, revealing complex patterns of vulnerability and adaptation across different ecological zones. The findings underscore the need for tailored, community-level interventions and integrated policy approaches to address the challenges posed by climate change to urban residential housing in Cameroon [30].

5 Concluding Remarks

Vulnerability of urban housing to the climate crisis in Cameroon reveals a complex interplay of environmental, socio-economic, and institutional factors across the country's diverse ecological zones. The study underscores the urgent need for tailored, community-level interventions and integrated policy approaches to address the multifaceted challenges posed by climate to urban residential housing. Key findings demonstrate that Cameroon's urban areas face varying climate-related risks, with coastal cities like Douala and Limbe contending with increased flooding and sea-level rise, while inland cities experience a range of issues from extreme rainfall events to heat stress. The study highlights the disproportionate impact on low-income households and informal settlements, echoing global trends in urban climate vulnerability.

The analysis of climatic trends reveals complex patterns, with some regions experiencing stable or increasing rainfall, while others face decreasing precipitation. This variability emphasizes the need for zone-specific adaptation strategies that consider local climate projections and socio-economic contexts. Socio-economic vulnerability emerges as a critical factor, with limited resources for housing retrofits and inadequate access to basic services exacerbating climate risks for many urban residents. The study points to a significant gap in community engagement and local institutional participation in vulnerability mapping and adaptation planning, highlighting the need for more inclusive and bottom-up approaches to urban climate resilience. Adaptation strategies and sustainable architectural designs show promise, with measures such as raised foundations and weather-resistant materials gaining traction. The adoption of these strategies varies across ecological zones, reflecting differing priorities and resource availability.

The integration of traditional knowledge and culturally appropriate strategies in adaptation efforts emerges as a valuable yet underutilized resource. Policy priorities identified in the study, particularly the emphasis on integrating climate adaptation into urban development policies, provide a roadmap for future action. The varying levels of stakeholder engagement across ecological zones underscore the need for targeted efforts to enhance participation, especially in more vulnerable regions.

As Cameroon continues to urbanize rapidly, the findings of this study provide crucial insights for policymakers, urban planners, and community leaders. By adopting a holistic, inclusive approach to urban climate resilience, Cameroon can work towards creating more sustainable, equitable, and climate-resilient cities for its growing urban population. These sustainable strategies will gain more impetus if they are integrated with Nature-based Climate Solutions and Ecosystem-based Climate Adaptation.

Acknowledgements We wish to thank the following persons who assisted us in the field: Dr Nsahalai Loveline Kongla, Dzebam Nelson Sahla, Suika Rita Nyuyfoni and Ako Tama Bawak

References

1. Andrews B, Carrillo-Silva D, Ilieva L, Kizhakkethottam C, Carulli E, Ang JC, Mutai J, Ware JP, Lanjouw N, King R, Atieno W (2023) The critical role of nature-based solutions for areas enhancing climate resilience in informal: an urban supplement to the UNFCCC Technical Guidelines on National Adaptation Plans. UN-Habitat, 67
2. Jamshed A, Patel C, Puriya A, Iqbal N, Rana IA, McMillan JM, Pandey R, Altaf S, Mehmood RT, bin Saad U (2024) Flood resilience assessment from the perspective of urban (in)formality in Surat, India: implications for sustainable development. Nat Hazards 120:9297–9326. https://doi.org/10.1007/s11069-023-06267-5
3. Williams AP, Abatzoglou JT, Gershunov A, Guzman-Morales J, Bishop DA, Balch JK, Lettenmaier DP (2019) Observed impacts of anthropogenic climate change on wildfire in California. Earth's Future 7(8):892–910
4. Williams DS, Máñez Costa M, Sutherland C, Celliers L, Scheffran J (2019) Vulnerability of informal settlements in the context of rapid urbanization and climate change. Environ Urbanisation 31(1):157–176
5. Tromeur E, Ménard R, Bailly JB, Soulié C (2012) Urban vulnerability and resilience within the context of climate change. Nat Hazard 12(5):1811–1821
6. Ngu NH, Tan NQ, Non DQ, Dinh NC, Nhi PTP (2023) Unveiling urban households' livelihood vulnerability to climate change: an intersectional analysis of Hue City, Vietnam. Environ Sustain Indic 19:100269
7. Rafael S, Martins H, Borrego C, Lopes M (2015) Urban vulnerability and resilience to climate change. WIT Trans Ecol Environ 198:379–390
8. Yonzoua CCF, Nanaa UJMP, Manjiaa MB, Pettanga C (2021) Mesure de la vulnérabilité des milieux urbains au Cameroun face au changement climatique. J Cameroon Acad Sci 17(2):115–129. https://doi.org/10.4314/jcas.v17i2.3
9. Di Ruocco A, Gasparini P, Weets G (2015) Urbanization and climate change in Africa: setting the scene. In Pauleit S, Kabisch S, Kombe WJ, Lindley S, Simonis I, Yeshitela K (eds) Urban vulnerability and climate change in Africa: a multidisciplinary approach. Future City. Springer, Cham, pp 1–35. https://doi.org/10.1007/978-3-319-03982-4_1
10. Tui S, Fakhruddin B (2022) Food for thought: Climate change risk and food (in)security in Tuvalu. Prog Disaster Sci 16:100255. https://doi.org/10.1016/j.pdisas.2022.100255
11. Tchouongsi EK (2021) Climate Variability and populations vulnerability to floods in urban spaces of southern countries: an analysis of determinants, threats and adaptation strategies in the city of Yaoundé in Cameroon. International J Innov Sci Res Technol 6(8):1329–1337
12. Bang DH, Miles L, Gordon R (2019) Evaluating local vulnerability and organizational resilience to frequent flooding in Africa: the case of Northern Cameroon. Foresight 21(2):266–284. https://doi.org/10.1108/FS-06-2018-0068
13 Tume SJP (2024) Rainfall reliability in the Bamenda Highlands and Coastal Lowlands of Cameroon: insights from Ndu and Douala (1957–2016). Environ Ecosyst Sci (EES) 8(1):16–25. https://doi.org/10.26480/ees.01.2024.16.25

14. Shabir F, Nain Md Z (2024) The economic impact of climate extreme events: a global perspective. Migration Lett 21(5), 1218–1234. www.migrationletters.com
15. Awazi NP, Kimengsi JN, Balgah RA, Forje WG (2023) Flood risks and community resilience across the five agroecological zones of Cameroon: reality and perspectives. In Fogwe ZN, Kimengsi JN, Sunjo TE, Kometa SS, Nkwemoh CA (eds) Celebrating ten years of restless rest in landmark research and human resource development: Festschrift for Professor Emeritus Cornelius Mbifung Lambi: development and environmental dynamics. The University of Bamenda Printing Press, pp 246–266
16. Balgah RA, Kimengsi JN, Ngwa KA, Nji T (2023) Floods management strategies across geo-ecological zones of Cameroon. In Fogwe ZN, Kimengsi JN, Sunjo TE, Kometa SS Nkwemoh CA (eds) Celebrating ten years of restless rest in landmark research and human resource development: Festschrift for professor Emeritus Cornelius Mbifung Lambi: development and environmental dynamics. The University of Bamenda Printing Press, pp 10–27
17. Awazi NP, Avana-Tientcheu ML, Alemagi D, Abanda FH, Enongene K, Nfornkah BN, Fobissie K (2024) Nature-based solutions for climate change adaptation and mitigation in Cameroon: realities and perspectives. In Filho WL, Nagy GJ, Ayal DY (eds) Handbook of nature-based solutions to mitigation and adaptation to climate change. Springer, Cham, pp 1–44. https://doi.org/10.1007/978-3-030-98067-2_77-1
18. Tume SJP, Awazi NP, Tantoh HB, Kunsom TK, Kongso ME, Fokeng RM, Bailack KM, Mairomi WH (2024) Climate variability in the Kilum-Ijim Mountain range of Cameroon: implications for rural livelihoods and food security. In Tripathi S, Bhadouria R, Garkoti SC (eds) Warming mountains: implications for livelihood and sustainability. Springer, Cham, pp 105–129. https://doi.org/10.1007/9783-031-62197-0_6
19. Tume SJP, Nyuyse YB, Foka SK, Kuyar C, Ngoran TL, Ngoran HA, Lueong LA, Ngala EN, Tambe AEM, Fanka IB, Kiseem BF, Mbinkar VJ (2024) Pathway for net zero policies: aligning nationally determined contributions (NDCS) for holistic sustainable development in Cameroon. Global Sci J 12(9):1279–1300
20. United Nations Framework Convention on Climate Change-UNFCCC (2021) Nationally Determined Contribution of the Republic of Cameroon. United Nations Framework Convention on Climate Change. UNFCCC, pp 1–20
21. Butler DR (ed) (2021) The Anthropocene, 1st ed, Routledge, London, 378. https://doi.org/10.4324/9781003208211
22. Iroume JY-A, Onguéné R, Koffi FD, Colmet-Daage A, Stieglitz T, Sone WE, Bogning S, Olinga JMO, Ntchantcho R, Ntonga J-C, Braun J-J, Briquet J-P, Etame J (2022) The 21st August 2020 flood in Douala (Cameroon): a major urban flood investigated with 2D HEC-RAS modeling. Water 14:1768. https://doi.org/10.3390/w14111768
23. Tume SJP, Kimengsi JN, Fogwe ZN (2019) Indigenous knowledge and farmer perceptions of climate and ecological changes in the Bamenda highlands of Cameroon: insights from the Bui Plateau. Climate 7(138):1–18. https://doi.org/10.3390/cli7120138
24. Douglas I, Anderson PML, Goode D, Houck MC, Maddox D, Nagendra H, Yok TP (eds) (2021) The Routledge handbook of urban ecology, 2nd edn. UN-Habitat and Routledge, London, p 1176
25. Gough KV, Yankson PWK, Wilby RL, Amankwaa EF, Abarike MA, Codjoe SNA, Griffiths PL, Kasei R, Kayaga S, Nabilse CK (2019) Vulnerability to extreme weather events in cities: implications for infrastructure and livelihoods. J Br Acad 7(s2), 155–181
26. Kabisch N, Stadler HKJ, Bonn A (eds) (2017) Nature-based solutions to climate change adaptation in urban areas: linkages between science, policy and practice, Springer, Cham
27. Girma Y, Terefe H, Pauleit S (2019) Urban green spaces use and management in rapidly urbanizing countries: the case of emerging towns of Oromia special zone surrounding Finfinne, Ethiopia. Urban Forestry Urban Greening 43:126357. https://doi.org/doi.org/10.1016/j.ufug.2019.05.019
28. Sygna L, Karen O'Brien K, Wolf J (eds) (2013) A changing environment for human security transformative approaches to research, policy and action, Routledge, London, 496. https://doi.org/10.4324/9780203109885

29. Awazi NP, Kimengsi JN, Balgah RA, Mairomi HW, Tume SJP, Tsufac AR (2024) Bioeconomy transition for the attainment of Cameroon's National Development Strategy (NDS30) goal of environmental and nature protection: assessing the all-encompassing contribution of agroforestry. In Singh K, Ribeiro MC, Calcioghi Ö (eds) Biodiversity and bioeconomy: status quo, challenges and opportunities, Elsevier, Amsterdam, pp 325–345
30. Africa Adaptation Initiative-AAI (2023) The Africa State of Adaptation Report 2023 (SOAR 23). European Union (EU), and United Nations Development Programme (UNDP), 189
31. Shackleton CM, Cilliers SS, Davoren E, du Toit MJ (2021) Urban ecology in the Global South: cities and nature, Routledge, London, 433. https://doi.org/10.1007/978-3-030-67650-6
32. Lueong LA, Tume SJP (2024) Institutional barriers to multi-sectorial climate change adaptation strategies in Cameroon. ISRG J Human Soc Sci, II I:14–19. https://doi.org/10.5281/zenodo.10466758
33. Lueong LA, Tume SJP (2024) Assessment of climate change financing mechanisms in Cameroon: a comprehensive review. J Geogr Environ Earth Sci Int 28(3):31–40. https://doi.org/10.9734/JGEESI/2024/v28i3754
34. Trisos CH, Adelekan IO, Totin E, Ayanlade A, Efitre J, Gemeda A, Kalaba K, Lennard C, Masao C, Mgaya Y, Ngaruiya G, Olago D, Simpson NP, Zakieldeen S (2022) Africa. In: Climate Change 2022: Impacts, Adaptation and Vulnerability. Contribution of Working Group II to the Sixth Assessment Report of the Intergovernmental Panel on Climate Change. Cambridge University Press, Cambridge, UK and New York, NY, USA, 1285–1455, https://doi.org/10.1017/9781009325844.011.
35. Burayidi MA, Allen A, Twigg J, Wamsler C (eds) (2019) The Routledge handbook of urban resilience, 1st ed, Routledge, London, 562. https://doi.org/10.4324/9780429506666

Role of Green Building Technologies in Achieving Climate-Resilient Cities

Bilal Hussain, Atiq ur Rehman, and Syed Asif Ali Naqvi

Abstract In the context of escalating climate change impacts, the role of green building technologies in fostering climate-resilient cities has become increasingly crucial. This chapter delves into the multifaceted aspects of green building technologies and their integral contribution to urban sustainability and resilience. It commences with exploring the fundamentals of green building technologies, including tracing their historical evolution, and investigating their core components, such as energy efficiency, waste reduction, indoor environmental quality, and the utilization of sustainable materials. The role of sustainable building materials is examined in terms of their environmental impact and contribution to building performance. The chapter further explores design strategies pivotal for climate resilience, focusing on passive architectural design, green roofing, and the incorporation of green spaces. The chapter concludes with a discussion of the diverse roles of green building technologies in enhancing urban resilience, emphasizing their necessity in the face of climate change.

Keywords Green building · Climate resilient cities · Energy efficiency · Sustainable materials · Urban sustainability

1 Introduction

Climate change has impacted cities with a heavy blow. The changing weather patterns ignited by high temperatures and recurrent calamities such as heat waves have increased the occurrence of heat islands in urban centers [1]. Climate change also aggravates income inequality due to displacement and health problems [2]. All the vital infrastructure in large cities, such as transport and water systems, are threatened and require expensive fixes to remain operational. Urban planning needs to incorporate climate adaptation for the benefit of all people in cities.

B. Hussain · A. ur Rehman · S. A. A. Naqvi (✉)
Department of Economics, Government College University, Faisalabad, Pakistan
e-mail: syedasif_1@yahoo.com

A. Cheshmehzangi et al. (eds.), *Designing Healthy Buildings and Communities*, Urban Sustainability, https://doi.org/10.1007/978-981-96-3057-8_5

Climate-resilient cities are planned and developed to cope with the physical and social impacts of climate change. These cities adopt green initiatives and climate-resilient structures to minimize risks and boost their readiness to deal with climate risks [3]. The focus of climate-resilient cities is safeguarding their people, structures, and finances against the negative impact of climate change. Incorporating green infrastructure, providing efficient energy systems, and embracing urban planning for sustainability lowers the urban heat island impact on cities. Climate-resilient cities also promote social justice by helping more vulnerable sectors of the population. Mitigating climate risks will help current and future generations to improve their overall quality of life and economic productivity. Figure 1 presents the conceptual framework of the study.

2 The Role of Green Building Technologies in Enhancing Urban Resilience

The resilience of cities can be significantly improved with the use of green building technologies, which improve the sustainability of buildings and lessen the negative effects of climate change. Examples of such technologies include energy-conserving systems, sustainable resources, and designs developed to reduce the consumption of resources and reduce greenhouse gas emissions [4]. Green spaces and tree plantations play a role in managing stormwater, reducing the urban heat island phenomena and improving air quality [5].

In addition to environmental value, structure and technology of green buildings also provides social and economic value. Efficient energy use in buildings reduces operational costs and enhances energy conservation, especially important in cases of disruption due to climatic variation [6]. Furthermore, these technologies can increase the thermal comfort of the building, which benefits the occupants' health [7]. Considering green building principles can establish optimum environments in terms of health and climate resilience. Table 1 compares climate-resilient and non-resilient cities based on energy efficiency, water management, green spaces, and other indicators.

3 Fundamentals of Green Building Technologies

Green building technologies pertain to aspects of construction that can be considered environment-friendly and cost-effective in every step of project. These technologies contribute towards minimizing the negative effects of human development on the environment, focusing on energy use in buildings and the use of sustainable materials [8]. In essence, green buildings are intended to reduce the use of energy, water, and other resources, generation of waste, and emission of pollutants, hence supporting

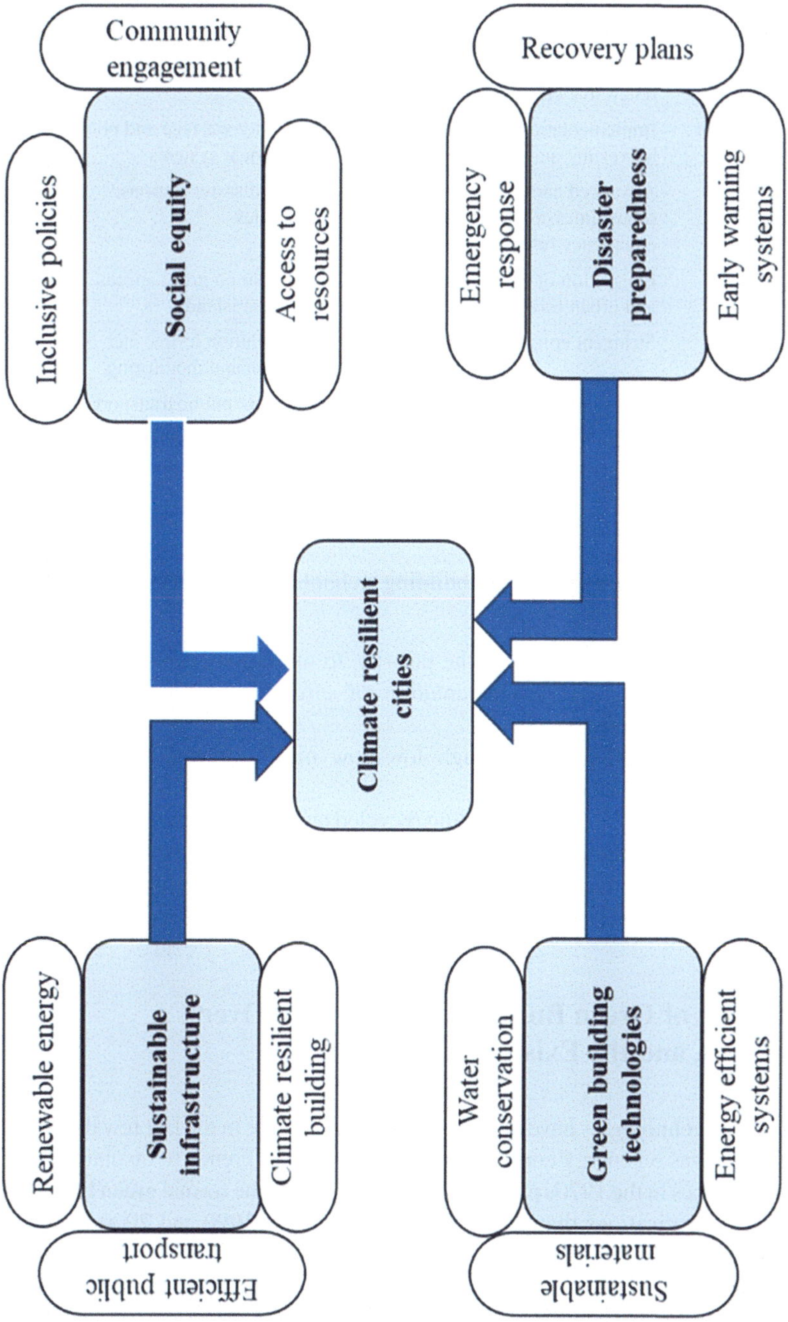

Fig. 1 Conceptual framework. *Source* Authors' own

Table 1 Comparison of climate-resilient vs non-resilient cities

Metric	Climate-resilient cities	Non-resilient cities
Energy efficiency	Use of energy-efficient technologies, smart grids and renewable energy	Reliance on non-renewable energy sources and outdated infrastructure
Water management	Implementation of rainwater harvesting and recycling	High water wastage and poor water management systems
Preparedness	Advanced early warning systems, robust infrastructure, and emergency response plans	Limited disaster response capabilities
Green spaces	Integration of parks, green roofs and urban forests	Limited or no green spaces, high urban heat islands
Air quality	Stringent emissions regulations	High pollution levels, lack of regulation and monitoring
Public transportation	Extensive and efficient public transport networks	Inadequate public transport systems
Waste management	Advanced recycling and waste reduction programs	High levels of waste generation, low recycling rates

environmental sustainability [9]. Green building technology is based on the following principles:

- Energy efficiency is considered as the priority, to minimize energy use through optimized insulation, and implementation of effective heating and cooling systems, as well as solar panels [4].
- Water efficiency is obtained through low-flow means of fixtures, rainwater collection, and greywater reuse [10].
- Sustainable materials are used, including recycled or domestic materials, to reduce environmental effects [8].
- Proper flow of fresh air, sufficient natural ventilation, and improved air quality provide healthy accommodation and working environments for occupants.

4 Evolution of Green Building Technologies Over the Years, and the Existing Trends

Green building technologies have undergone a drastic change in the last few decades. First, the focus was on energy conservation, with special reference to insolation and efficient appliances in the 1970s prompted by the oil crisis. The formal green building standards and certifications that started emerging between 1990 and 2000 include, among others, the Building Research Establishment Environmental Assessment Method (BREEAM), which offered guidelines for green building practices. More recently the focus has been on multi-dimensional solutions that encompass water use, materials, energy literacy, and broader sustainability [10]. Smart building technologies such as the Internet of Things (IoT) have been incorporated into building management systems, thus allowing real-time control and analysis of energy consumption and comfort of occupants in a building. Also, an increasing number of constructed buildings conform to the requirements of net-zero energy buildings, which utilize renewable energy technologies like solar and wind to generate energy equivalent to their consumption [11]. Current trends in green building technologies are listed below.

- **Smart Building Systems:** Integration of IoT for automated energy management, lighting, and climate control
- **Biophilic Design:** Incorporation of natural elements to improve occupant well-being and reduce stress
- **Advanced Insulation Materials:** Use of high-performance insulation to enhance energy efficiency
- **Renewable Energy Integration:** Integration of solar panels, wind turbines, and other renewable sources into building design
- **Water Recycling Systems:** Systems for grey water recycling and rainwater harvesting
- **Sustainable Building Materials:** Use of recycled, renewable, and locally sourced materials to minimize environmental impact
- **Passive Solar Design:** Architectural design that maximizes natural light and heat retention
- **Energy Storage Solutions:** Integration of batteries and other energy storage technologies for energy resilience

5 Energy-Efficient Technologies

5.1 Insulation Materials and Techniques

Energy conservation is one of the pillars of green building systems, and insulation is the key element in energy conservation. Insulation materials and procedures have undergone adaptations that enhance their insulative characteristics, reduce heat transfer, and maintain the internal environment at optimal temperatures. Recently, various types of insulation have been incorporated to improve performance, including spray foam, rigid foam board, and modern fiberglass. Modern methods include radiant barriers and reflective insulation that keep buildings cooler. Structures made of structural insulated panels (SIPs) are efficient at keeping heat in as well as improving the structural value of a space [12]. Also, green roofs and walls, which have vegetation layers, provide insulation and decrease the urban heat island effect [5]. These materials and techniques greatly reduce heating and cooling requirements and in general provide a more sustainable building.

5.2 *Renewable Energy Integration*

Integrated renewable energy systems include solar panels and wind turbines installed on buildings' façades. Solar panels generate power from sunlight, supporting the reduction of fossil fuels as a source of power. Photovoltaic technology has improved over the years in terms of efficiency and cost, which makes solar panels suitable for use in residences, businesses, and industrial applications [13]. The implementation of building-integrated photovoltaics, particularly for roofs, façades, and windows, is useful as it preserves the aesthetic value of the solar energy system and means they can be installed without requiring additional space [14]. Wind turbines can also be integrated into buildings, but these are less common due to space limitations and appearance issues.

5.3 *Smart Energy Systems for Energy Management*

In smart buildings, technology that boosts energy and building effectiveness is incorporated into the infrastructure. These systems deploy sensors, IoT devices, and artificial intelligence, allowing real-time supervision of energy consumption and management [15]. The smart systems, which learn from their building's occupants, work to maintain a balanced temperature that only requires a small amount of energy [16]. Building automation systems (BAS) integrate various systems in a green building, including heating, ventilation, air conditioning, and lighting, for smart energy management. These systems have numerous functional capabilities, including turning off the lights automatically, ensuring maintenance to enhance the efficiency of the equipment, and identifying faults to avoid energy waste [17]. Furthermore, the integration of smart grids provides communication with the power grid to control, manipulate, and appropriately coordinate load demand and supply [18]. In addition to improving energy efficiency, these technologies also promote general sustainability and versatility.

6 Sustainable Building Material

Sustainable materials curtail the negative impacts of construction on the environment and they are manufactured in ways that promote the least exploitation of resources and conform to sustainability criteria with a lesser effect on the climate. Recycled materials reuse postindustrial or post-customer waste and convert those into new construction products. For example, recycled steel requires 75% less energy to produce than new steel, thus cutting down on greenhouse gases. Recycled concrete aggregate (RCA) is another common type; it involves using crushed concrete from buildings and other structures and reusing the material as aggregate in new concrete, avoiding virgin material and eliminating waste that would otherwise go to landfill [19].

Renewable materials can reproduce on their own over time. For example, bamboo grows much faster than traditional timber; the annual increment of bamboo ranges between 3–5 years, while that of most hardwoods is 20–50 years. It is also harder, more flexible, and can be used for flooring, structures, and finishes. Locally sourced materials help minimize the carbon footprint from transportation and help boost local industries. For instance, quarrying stone or harvesting timber from the local region cuts down on emissions from transportation and encourages local industries [20].

6.1 Lifecycle Assessment of Building Materials

Lifecycle assessment of building materials is a method used to gauge the lifelong impacts of construction materials on the environment. This approach analyzes the chances of lowering the impact of construction on the environment and encourages the use of materials that are friendlier to the planet. The first part of a building material's lifecycle is the extraction of raw materials and the transformation of these into useful products. These processes can have major negative bearings on the environment, such as through resource exploitation, destruction of habitats, and high energy use. For instance, cement contributes 8% of global CO_2 emissions [21]. Reducing the adverse effects of cement production can be achieved by using other materials, including fly ash or slag [22].

In the use phase, the durability and energy intensity of a building is fortified by the qualities of the materials that have been used. Building elements with high thermal thickness, and dense materials including concrete and bricks, can work as low energy-consuming interiors as they require little energy for heating and cooling. However, there is also a need to consider maintenance requirements, alongside potentially harmful emissions, like volatile organic compounds, that come from specific paints and finishes. Demolition of the building is the last stage of the lifecycle, including the disposal and reuse of materials. This stage is crucial as far as environmental effects are concerned. There is no limit to recycling steel and aluminum, which have the ability to retain their characteristics; however, some materials, like plastics and composites, may be difficult to recycle and may end up increasing landfill waste. Thus, the stakeholders can decide on the materials to use by looking not only at the monetary costs of purchase and efficiency, but also at the other costs that are incurred in the provision and disposal of materials. The lifecycle assessment method also promotes the application of improved strategies and products that have a reduced effect on the environment.

6.2 Impact of Sustainable Materials on a Building's Performance and Environmental Footprint

The use of sustainable materials improves insulation and greatly enhances a building's thermal performance, durability, and energy efficiency. For instance, heavy constructions like rammed earth and concrete walls are favourable because they retain heat for a long period, and therefore, little artificial heating and cooling is required. Also, the use of paints and finishes with low volatile organic compounds improves indoor air quality because the emission of hazardous chemicals is reduced. Moreover, the use of sustainable materials reduces a building's ecological footprint in many ways:

- **Lower embodied energy:** Some materials, like bamboo, cork, and recycled steel, have less embodied energy than other materials because less energy is used in the making process and during transportation. For instance, bamboo possesses less embodied energy and grows faster than conventional hardwoods and is hence considered better for the environment.
- **Reduced greenhouse gas emissions:** Regular use of green materials in production can greatly reduce greenhouse gases. For instance, recycled steelmaking emits 75% less CO_2 than that the production of new steel from scratch. Likewise, the incorporation of fly ash or slag in the concrete lowers the CO_2 emissions relative to the production of cement [22].
- **Resource conservation:** A common way that sustainable materials are incorporated into a building is through the use of recycled and renewable resources, with greatly reduced use of virgin materials. For instance, recycled concrete aggregate is made from concrete wastes from demolished buildings and structures; using this helps to conserve natural aggregates and minimize landfilling [19]. Substances such as cork and bamboo are renewable materials that can be replaced within a short period, which offers a solution to scarce resources.

7 Design Strategies for Climate Resilience

Different design strategies for climate resilience are given in this section.

7.1 Passive Architectural Design

By utilizing the natural environment, passive architectural design reduces dependence on excessive mechanical systems to lower energy consumption, and also enhances the comfort and efficiency of the building. Passive design elements include natural ventilation and daylighting. Natural ventilation is a passive method of air movement through buildings by the combined effects of wind and thermal buoyancy; it leads to the conservation of energy while at the same time enhancing the quality of indoor air. For example, cross-ventilation techniques, in which openings such as windows or vents are located on opposite sides of a room, greatly improve airflow. Use of natural ventilation lessens cooling energy use by as much as 60% in moderate climates [23]. Other strategies of natural ventilation include stack ventilation, and atriums and ventilation shafts. These methods improve the flow of air to keep internal environments at appropriate temperatures, and enhance the internal circulation of air.

Daylighting makes use of natural light at its optimum, to eliminate the use of artificial light, improve occupant comfort, and save on costs. Adequately designed daylighting helps to save energy use by up to 40% [24]. Further, for controlling artificial lighting during the day, light shelves and reflectors that spread natural light over most of the interior are incorporated.

7.2 Green Roofing and Wall Systems

Green roofing and wall systems are a sophisticated application of technology in the construction of buildings, involving the use of vegetative layers on the roof and facades. They elevate the level of resistance against temperature for buildings and assist in the framework for sustainable urbanism. They naturally control the temperature in the dwelling to the extent that in many cases, the occupants rarely need heating or cooling systems. Green roofs can lower the intensity of energy utilization in air conditioning by at least 75% in summer. Stormwater is also retained by green roofs, hence controlling the instances of flooding, especially in urban areas. By providing shading and evaporative cooling through a process known as evapotranspiration, green roofs can cool the air around them and reduce the intensity of the urban heat island.

Green walls, also known as living walls, are vertical surfaces on which plants are grown either directly into the soil or in hanging modular units fixed to the wall, indoors or outdoors. The systems provide numerous environmental gains: They assist in controlling the temperature within a building by offering functions of insulation and shading; plants on the green walls can help purify polluted city air; and they work well in the removal of particulate matter and volatile organic compounds (VOC). Green infrastructure adds to the beauty of buildings and other urban structures, which has been noted to have a positive impact on the mental health of residents.

7.3 Building Orientation and Layout for Energy Efficiency

Building orientation and layout are important elements in designing energy-efficient buildings. The orientation of the houses allows maximum winter solar gain. Global attention towards energy conservation has grown to reduce CO_2. With reference to building energy performance, wisdom of using earth as temperature moderator against tough weather has extraordinary potential to become a solid solution against the energy inefficiency of heating [25]. Orientation, proper designing and architects can optimize natural light, ventilation, and thermal performance to reduce energy consumption for heating, cooling, and lighting.

7.4 Green Spaces and Biodiversity

When it comes to improving the resilience and sustainability of cities, implementing green spaces and supporting biodiversity are of special importance. Green spaces such as parks, gardens, and green corridors serve a variety of ecological functions and also provide social value and economic returns; among these benefits, the promotion of green health and climate-resilient cities are significant. Green areas are sources of food for small animals and thus, incorporating green areas into urban environments supports ecological balance. Urban regions with spatial reserves that offer good possibilities for integrating precisely developed green areas can accommodate a large number of plant species as well as animals, including local and some migrating species. This is important because birds need trees in urban parks and gardens to feed and breed, and they play an important role in pest control and pollination. Research has indicated that the integration of greenery in the urban setting can boost the biodiversity in the region by as much as 50%, improving the functionality of the ecosystems through services like removal of air and water pollutants, absorption of carbon, and moderation of temperature.

Green spaces are involved in reducing the urban heat island effect, managing stormwater, and improving air quality, which are all aspects of climate change adaptation. Through shading and evapotranspiration, vegetation can decrease urban thermal conditions by as much as 2–3 °C, therefore reducing the need for air conditioning and in turn, reducing energy use. Green spaces also provide rainwater infiltration services to the community and help to minimize flooding. Importantly, green spaces contribute to the qualitative characteristics of the urban environment by providing opportunities for recreation and fostering social inclusion, which boost the physical and psychological well-being of the population. Research shows that the presence and use of green space reduces stress, anxiety, and depressive symptoms, which make for improved health of the community. From an economic perspective, green areas can increase property values and tourism, bringing in revenue to the local economy or government. Urban parks can increase property values in the surrounding area by as much as 20%, primarily due to a general desire to live in a healthier environment. Moreover, urban green spaces are of critical importance to the sustainable development of urban areas, in terms of ecological climate and economic value. Therefore, cities that integrate green spaces and improved biodiversity are better placed to set healthy environments for their people.

8 Conclusions

Green and smart building technologies are contributing enormously to the resilience of climate-vulnerable cities. Green buildings are mitigating the devastating effects of climate change with energy-efficient technologies, renewable energy integration, water and waste management, and sustainable materials. The use of energy-efficient technologies, which include heating, ventilation, and air conditioning (HVAC) systems, heating-resistant insulation materials, and integration of renewable solar and wind energy, decreases the strain on already scarce energy resources and dampens the use of fossil fuels. In addition, smart energy management systems incorporate building automated systems (BAS) and IoT to efficiently control and manage energy consumption using real-time data.

Green buildings use sustainable materials that are reused and recycled. Materials that are more sustainable improve the insulation of a structure and greatly enhance its thermal performance and durability, also reducing energy use. Passive architectural designs, like natural ventilation and daylighting, are also important as they conserve energy as well as improve the quality of indoor air. Green roofs and wall systems contribute by helping to maintain indoor temperature without the need for air conditioning and other mechanical sources, while green spaces in cities offer several ecological benefits. Building orientation is also a key factor to consider, as part of building design and layout that supports the efficient use of solar and wind energy in both summer and winter. The use of water-efficient appliances and fixtures in buildings can greatly save water for sustainable use. It is clear that there are many sustainable practices at our fingertips—implement these practices will help to combat climate change and make cities more resilient.

References

1. Huang G, Zhou W, Cadenasso ML (2021) Is everyone hot in the city? spatial pattern of land surface temperatures, land cover and neighborhood socioeconomic characteristics in Baltimore, MD. J Environ Manage 259:109691
2. Sharifi A, Simangan D (2021) Climate resilience and urban sustainability: integration of these paradigms in urban planning. Sustainability 13(2):591
3. Meerow S, Newell JP, Stults M (2016) Defining urban resilience: a review. Landsc Urban Plan 147:38–49
4. Berardi U (2016) A cross-country comparison of the building energy consumptions and their trends. Resour Conserv Recycl 123:230–241
5. Sodoudi S, Shahmohamadi P, Vollack K, Cubasch U, Che-Ani AI (2018) Mitigating the urban heat island effect in megacity Tehran. Adv Meteorol 2018:1–12
6. Husain A, Cohan DS (2021) The environmental and economic impacts of increasing renewable energy deployment in the United States. Energy Policy 148:111956
7. Fisk WJ (2017) The ventilation problem in schools: literature review. Indoor Air 27(6):1039–1051
8. Kibert CJ (2016) Sustainable construction: green building design and delivery. Wiley
9. U.S. Green Building Council (2020) LEED v4.1 Reference Guide for Building Design and Construction

10. Jiang X, Wong LT, Yang L (2021) Water conservation in buildings: an overview of technologies and strategies. Water 13(3):371
11. Attia S, Hamdy M, O'Brien W, Carlucci S (2021) Assessing gaps and needs for integrating building performance optimization tools in net zero energy buildings design. Energy Build 224:110294
12. Biswas K, Shrestha SS, Bhandari M, Desjarlais AO (2021) A review of advancements in building insulation technologies and their applications. Renew Sustain Energy Rev 142:110877
13. International Renewable Energy Agency (IRENA) (2021) Renewable Power Generation Costs in 2020. IRENA
14. Shukla AK, Sudhakar K, Baredar P (2018) A comprehensive review on design of building integrated photovoltaic system. Energy Build 128:99–110
15. Aste N, Adhikari RS, Buzzetti M (2017) Smart building integration in the urban context. Energy Build 154:336–348
16. Yang R, Zhang K (2020) Design and implementation of an intelligent HVAC control system for energy-efficient buildings. Appl Energy 269:114928
17. Dounis AI (2017) Artificial intelligence for energy conservation in buildings. Adv Build Energy Res 11(1):79–91
18. Mocnej J, Kadera P, Musil P, Fiedler P, Strelec J (2020) Real-time building energy management: a case study on the impact of IoT and big data technologies in the optimization of heating system energy consumption. Procedia Comput Sci 167:2405–2414
19. Tam VWY, Gao XF, Tam CM, Ng KM (2021) Recycled concrete aggregates in new concrete: sustainability assessment. J Clean Prod 278:123689
20. Jones P (2020) Local materials: the role of locally sourced materials in sustainable construction. Sustain Constr J 12(3):58–67
21. Lehne J, Preston F (2018) Making concrete change: innovation in Low-carbon cement and concrete, Chatham House
22. Scrivener KL, John VM, Gartner EM (2018) Eco-efficient cements: potential economically viable solutions for a low-CO_2 cement-based materials industry. United Nations Environment Programme
23. Oropeza-Perez I, Ostergaard PA (2014) Energy saving potential of utilizing natural ventilation under warm conditions–a case study of Mexico. Appl Energy 130:20–32
24. Ahmad RM, Reffat RM (2018) A comparative study of various daylighting systems in office buildings for improving energy efficiency in Egypt. J Build Eng 18:360–376
25. Alkaff SA, Sim SC, Efzan ME (2016) A review of underground building towards thermal energy efficiency and sustainable development. Renew Sustain Energy Rev 60:692–713

Healthy Community Interventions

Enhancing Neighborhood Food Availability for Safer and Healthier Urban Environments: A Cross-Sectional Investigation in Changsha, China

Ke Peng, Zhimou Peng, and Rongpeng Zhang

Abstract This chapter embarks on a pivotal exploration into the enhancement of neighborhood food availability, recognizing its profound importance in fostering safer and healthier urban environments. Access to nutritious food within neighborhoods is not only a fundamental aspect of public health but also plays a crucial role in shaping the social, economic, and environmental fabric of urban communities. In the wake of the COVID-19 pandemic, which has magnified existing disparities in food access, understanding and addressing challenges related to neighborhood food availability have become increasingly urgent. The ability of residents to access a diverse range of affordable and nutritious food options within their immediate surroundings directly impacts their dietary choices, health outcomes, and overall well-being. Moreover, disparities in food access are often intertwined with broader issues of social equity and justice, with marginalized communities disproportionately affected by limited access to healthy food options. Addressing these disparities requires a multifaceted approach that considers factors such as neighborhood infrastructure, socio-economic status, and food retailing practices. Against this backdrop, our study focuses on Changsha, China, as a case study to explore the complex dynamics of neighborhood food availability. By examining the relationship between neighborhood housing prices and food outlet availability, we aim to uncover insights that can inform strategies for promoting equitable access to nutritious foods and improving public health outcomes. Additionally, we investigate the perception and use of daily consumer facilities by the elderly in suburban areas, assessing how factors like distance and facility size influence their accessibility and utilization. Lastly, we explore the correlation between neighborhood food environments and cardiovascular and metabolic health indicators, with a focus on the mediating effect of commuting distances. Through this comprehensive exploration, we seek to contribute to a deeper understanding of the factors influencing neighborhood food availability and to inform evidence-based interventions aimed at creating safer, healthier, and more inclusive

K. Peng · Z. Peng · R. Zhang (✉)
School of Architecture and Planning, Hunan University, Changsha, P.R. China
e-mail: zhangrongpeng@hnu.edu.cn

A. Cheshmehzangi et al. (eds.), *Designing Healthy Buildings and Communities*, Urban Sustainability, https://doi.org/10.1007/978-981-96-3057-8_6

urban environments for all residents. By prioritizing food availability as a key component of urban planning and public health efforts, we endeavor to promote equity, resilience, and well-being within urban communities.

Keywords Food availability · Health promotion · Resilience · Well-being · Built environment

1 Introduction: Addressing Food Availability for Urban Safety and Health

1.1 Definition of Food Availability

Urban environments play a crucial role in shaping the health and well-being of their residents. The design, infrastructure, and resources available within a city can significantly influence lifestyle choices, access to services, and overall quality of life. In the pursuit of healthier urban environments, one critical aspect that requires attention is the availability of nutritious food. Food availability refers to the adequacy of the supply of healthy food; examples in the food environment might include the presence of certain types of food stores or restaurants near people's homes [1]. This chapter delves into the intricate relationship between neighborhood food availability and urban health, highlighting why this topic is essential for creating safer and healthier urban environments.

1.2 Neighborhood Food Availability and Public Health

Access to nutritious food within neighbourhoods is a fundamental aspect of public health [2]. It significantly influences dietary behaviors and health outcomes, contributing to the prevention of chronic diseases such as cardiovascular diseases (CVD), diabetes, and obesity [3]. Adequate food availability ensures that residents can make healthy dietary choices, which are essential for maintaining optimal health and preventing the onset of nutrition-related diseases. Poor food availability in urban neighborhoods often leads to higher consumption of unhealthy, processed foods, which are linked to an increased prevalence of chronic diseases such as cardiovascular diseases, diabetes, and obesity [4]. Therefore, ensuring adequate food availability is crucial for fostering healthier urban populations and reducing the burden of diet-related illnesses.

Cardiovascular Diseases (CVD): Research indicates that neighborhoods with limited access to healthy food options are associated with higher rates of cardiovascular diseases. For instance, studies conducted in the United States and the United

Kingdom have shown that residents of food deserts—areas with poor access to affordable, healthy food—exhibit higher rates of hypertension and heart disease [5, 6]. Interventions aimed at increasing the availability of fresh fruits and vegetables in these areas have been linked to improvements in heart health and reductions in CVD risk factors.

Diabetes: The prevalence of type 2 diabetes is closely linked to dietary patterns influenced by food availability. Studies in both developed and developing countries have found that limited access to nutritious food is associated with higher rates of diabetes [7, 8]. In urban areas of South Asia and Africa, where fast food outlets are more accessible than fresh produce markets, there has been a notable increase in diabetes incidence. Programs that improve access to healthy food options and promote dietary education have shown promising results in reducing diabetes rates.

Obesity: Obesity is another major health concern linked to food environments. Research from Australia, Canada, and various European countries indicates that neighborhoods with high densities of fast-food restaurants and convenience stores have higher obesity rates [9–11]. Conversely, areas with greater access to supermarkets and farmers' markets tend to have lower obesity prevalence. Policy measures, such as zoning laws to limit fast food outlets and incentives for healthy food retailers, have been effective in addressing obesity at the community level.

Cancers: Certain cancers, particularly those of the digestive system, are associated with diet quality and food availability. Studies from Japan and France have highlighted the role of healthy diets rich in fruits, vegetables, and fiber in reducing the risk of colorectal and stomach cancers [12]. Conversely, diets high in processed meats and low in plant-based foods are linked to increased cancer risk. Ensuring access to healthy food can thus play a crucial role in cancer prevention.

Mental Health: Emerging research also suggests a connection between food availability and mental health. Studies from Nordic countries and the Mediterranean region indicate that diets rich in whole foods, such as fruits, vegetables, whole grains, and fish, are associated with lower rates of depression and anxiety [13, 14]. In contrast, diets high in processed foods and sugars are linked to poorer mental health outcomes. Initiatives to improve food environments can contribute to better mental health and overall well-being.

1.3 Impact of Neighborhood Food Availability on Social Equity and Community Well-Being

Access to nutritious food can enhance mood, cognitive function, and social interactions, thereby fostering a more vibrant and cohesive community [15]. Neighborhood food availability also plays a significant role in shaping the social, economic, and environmental fabric of urban communities [16]. Marginalized communities, including low-income and minority populations, frequently suffer from inadequate access to nutritious food due to systemic inequalities. These communities are often situated in food deserts, where healthy food options are scarce, or food swamps, where unhealthy options dominate. Such disparities exacerbate existing health inequities, as individuals in these areas face higher risks of developing chronic diseases. Addressing food availability is thus essential for promoting social equity, ensuring that all residents, regardless of socio-economic status, have equal access to healthy food and the associated health benefits. Further, access to nutritious food supports local economies by encouraging the growth of food markets and grocery stores. It fosters social cohesion by providing communal spaces where residents can gather and interact. Additionally, promoting local food systems can have positive environmental impacts by reducing food miles and supporting sustainable agricultural practices. Therefore, improving food availability is not only a health imperative but also a vital component of creating vibrant and sustainable urban communities. By integrating food availability into urban planning, cities can promote economic development, environmental sustainability, and social inclusivity, making urban areas more livable and resilient.

In a nutshell, enhancing food availability has profound implications for community well-being and social cohesion. Access to nutritious food fosters a sense of security and well-being among residents, contributing to stronger, more resilient communities. When communities have reliable access to healthy food, it not only supports physical health but also enhances social interactions and community bonds. Initiatives such as farmers' markets, community gardens, and local food cooperatives can serve as community hubs, bringing people together and fostering a sense of belonging. By improving food availability, urban planners and policymakers can support the holistic well-being of urban populations, making cities healthier and more inclusive places to live.

1.4 Lessons Learned and Best Practices for Improving Food Availability

The international body of research underscores the importance of holistic approaches to improving food availability. Successful interventions often involve multi-sector collaboration, combining efforts from public health, urban planning, and community organizations. Best practices include implementing policies that support local food systems, enhancing transportation networks to improve food access, and engaging community members in the planning and implementation of food-related programs.

Governments worldwide have adopted various policy interventions to improve food availability. Examples include Mexico's tax on sugary beverages, which has led to a decline in soda consumption and an increase in water intake, and Brazil's school feeding programs that prioritize local, nutritious foods. These policies highlight the role of governmental action in shaping healthier food environments. Community-driven initiatives, such as urban gardens, farmers' markets, and food co-operatives, have proven effective in enhancing food availability. These solutions not only improve access to healthy foods but also foster community engagement and social cohesion. Programs like Detroit's urban farming movement and Nairobi's community garden projects serve as models for other cities aiming to improve food environments through grassroots efforts.

1.5 Relevance in the Context of the COVID-19 Pandemic

The COVID-19 pandemic has underscored the critical importance of food availability within urban environments. As cities faced lockdowns and movement restrictions, the accessibility of food became a paramount concern. Disruptions in supply chains, panic buying, and changes in consumer behavior highlighted vulnerabilities in urban food systems. Neighborhoods with limited food outlets experienced severe difficulties in accessing nutritious food, exacerbating food insecurity and health disparities. This period revealed the urgent need for resilient and accessible food systems that can withstand crises and ensure continuous access to healthy food for all residents.

The pandemic has disproportionately affected vulnerable populations, amplifying existing social and economic disparities related to food availability. Low-income households, already struggling with limited access to nutritious food, faced additional challenges due to job losses, reduced incomes, and increased reliance on food assistance programs. Marginalized communities, often located in food deserts or swamps, found it even harder to obtain healthy food during the pandemic. This exacerbation of food insecurity among vulnerable groups highlighted the necessity of addressing social inequities in food access as part of broader public health strategies. The COVID-19 pandemic provides valuable lessons for urban planning and public health, emphasizing the need for robust food systems that promote health equity and resilience. Cities must prioritize the integration of food availability into urban planning processes, ensuring that all neighborhoods have access to a diverse range of nutritious food options. This involves strategic placement of food outlets, support for local food production, and policies that encourage healthy eating habits. By learning from the pandemic, urban planners and policymakers can develop more resilient, equitable, and health-focused urban environments that better serve their populations in both normal and crises.

2 Conceptual Framework: Food Availability and Healthy Urban Planning

2.1 *Link Between Food Availability and Dietary Behaviors and Chronic Diseases*

Dietary patterns characterized by high intake of sugar, salt, and fat are prevalent in areas with limited access to healthy food options. Fast-food restaurants and convenience stores, which are often more accessible in these areas, typically offer foods that are calorie-dense but nutrient-poor. These dietary patterns contribute to the development of obesity, insulin resistance, hypertension, and other metabolic disorders (Fig. 1). The proliferation of unhealthy food outlets in certain neighborhoods creates an environment where unhealthy eating is the norm, exacerbating public health issues [17]. Addressing these food environment disparities is crucial for promoting healthier eating habits and reducing the incidence of diet-related chronic diseases. Conversely, diets rich in fruits, vegetables, and legumes are associated with numerous health benefits, including lower risks of CVD, diabetes, and obesity. These foods are high in essential nutrients, fiber, and antioxidants, which support overall health and protect against chronic diseases. Neighborhoods with better access to supermarkets and farmers' markets, where these protective foods are readily available, show higher consumption rates of these health-promoting foods [18]. Encouraging the availability and consumption of fruits, vegetables, and legumes through urban planning and policy interventions can lead to substantial improvements in public health outcomes.

2.2 *Link Between Food Availability and Built Environment*

The built environment encompasses the physical layout and infrastructure of urban areas, including transportation systems, land use patterns, and the location of food outlets. Food availability within the built environment is influenced by factors such as the distance to food stores, the availability of public transportation, and the walkability of neighborhoods. Urban planning plays a crucial role in shaping the built environment to improve food availability. For instance, creating mixed-use neighborhoods with a blend of residential, commercial, and recreational areas can enhance access to healthy food options.

Urbanization significantly transforms food systems, impacting both the availability and quality of food in urban areas. As cities expand, agricultural land often gives way to urban development, reducing local food production and increasing reliance on imported foods. This shift can lead to a decrease in the availability of fresh, locally sourced produce and an increase in processed and packaged foods that are easier to transport and store. Additionally, urban sprawl can create physical

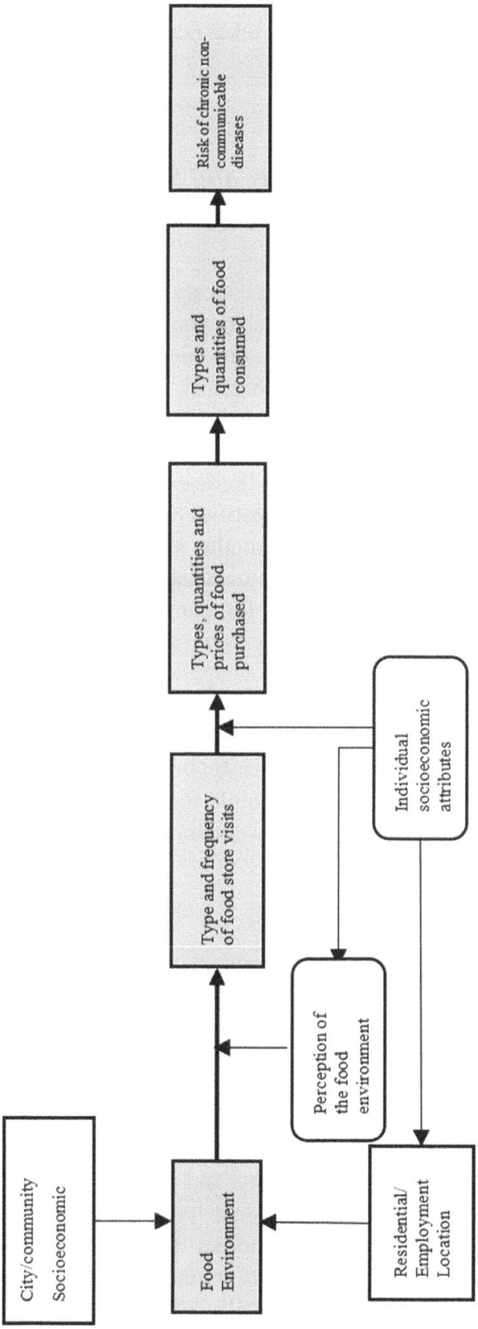

Fig. 1 The link between food environment and risk of chronic non-communicable diseases. *Source* Drawn by Authors

barriers to accessing food, as supermarkets and grocery stores may be located far from residential areas, particularly in low-income neighborhoods. Urban planners must consider these dynamics to ensure that urbanization does not compromise food availability and public health.

2.3 Link Between Food Availability and Socio-Economic and Demographic Factors

Socio-economic factors such as income and education levels play a significant role in shaping food environments and dietary behaviors. Lower-income neighborhoods often have fewer supermarkets and more convenience stores, limiting access to affordable, healthy food options [19]. Additionally, individuals with lower education levels may have less nutritional knowledge, further impacting their dietary choices. Economic constraints and limited knowledge create barriers to healthy eating, contributing to higher rates of chronic diseases in disadvantaged communities. Addressing these socio-economic disparities through targeted interventions is essential for improving food access and promoting health equity.

Demographic factors, including age and occupation, also influence food availability and dietary behaviors. Older adults, for example, may face mobility issues that limit their access to food outlets, while younger, working adults may opt for convenience foods due to time constraints. Occupation-related factors, such as long working hours and shift work, can also impact food choices and eating patterns. Tailoring interventions to address the specific needs of different demographic groups can enhance the effectiveness of strategies aimed at improving food environments and health outcomes. For instance, providing mobile food markets or delivery services can help elderly residents, while workplace wellness programs can support healthier eating among working adults.

2.4 Theoretical Underpinnings and Models

2.4.1 Socio-Ecological Framework

The socio-ecological framework is a foundational model for understanding food availability and its impact on public health. This framework emphasizes the interplay between individual, interpersonal, community, and societal factors in shaping food access and dietary behaviors. At the individual level, factors such as knowledge, attitudes, and preferences influence food choices. Interpersonal factors, including family and social networks, can either support or hinder healthy eating habits. At the community level, the availability and accessibility of food outlets, as well as cultural and social norms, play significant roles. Finally, societal factors, including policies,

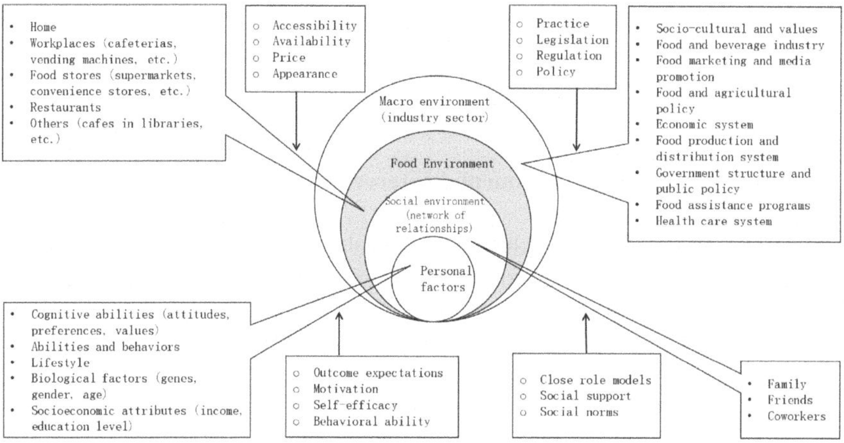

Fig. 2 Factoring affecting dietary quality based on socio-ecological models. *Source* The author modified the drawing according to the paper of Story et al. and combined it with the characteristics of the domestic food environment

economic conditions, and media influences, shape the broader food environment (Fig. 2).

Socio-ecological models provide a framework for understanding the complex interplay between urban environments and food availability. These models consider how various layers of influence, from individual behaviors to broader environmental and policy contexts, interact to shape food access. In urban settings, factors such as residential density, land use patterns, and transportation infrastructure play critical roles in determining how easily residents can access nutritious food. High-density, mixed-use neighborhoods tend to offer better food access due to the proximity of food outlets and the availability of public transportation. Conversely, low-density, car-dependent suburbs may present significant barriers to food access, particularly for individuals without reliable transportation. Socio-ecological models highlight the need for integrated urban planning approaches that address multiple levels of influence to improve food availability [20].

2.4.2 Multi-Level Models in Food Availability Research

Multi-level models are increasingly used in food availability research to capture the complex interactions between different levels of influence. These models allow researchers to examine how individual, community and societal factors interact to shape food access and dietary behaviors. For instance, a multi-level model might explore how individual factors such as income and education interact with community factors like the density of food outlets and societal factors such as food policies. By capturing these interactions, multi-level models provide a more nuanced understanding of the determinants of food availability and identify key leverage points for intervention. This approach is particularly useful in urban settings, where the interplay between individual, community, and societal factors can be highly complex and dynamic.

2.4.3 Behavioral and Environmental Interventions

Behavioral interventions aim to change individual and community dietary behaviors through education, motivation, and support. These interventions often include nutrition education programs, social marketing campaigns, and initiatives that promote healthy eating habits. Environmental interventions, on the other hand, focus on modifying the physical and policy environments to improve food access. Examples include urban planning strategies that increase the number and diversity of food outlets, policies that provide incentives for grocery stores to locate in underserved areas, and regulations that limit the density of fast-food outlets. Combining behavioral and environmental approaches can create environments that not only promote healthy eating behaviors but also make it easier for individuals and communities to access nutritious foods. This integrated approach is essential for addressing the complex challenge of food availability in urban settings.

2.5 Urban Planning Tools and Policy Interventions

Effective urban planning and policy interventions are essential for enhancing food availability in urban environments. Zoning regulations can be used to encourage the establishment of grocery stores and farmers' markets in underserved areas, while restrictions on fast food outlets can help reduce the prevalence of unhealthy food options. Transportation policies that improve public transit access to food outlets can also play a crucial role in enhancing food availability. Additionally, urban agriculture initiatives, such as community gardens and rooftop farms, can increase the supply of fresh, locally-grown produce in cities. Policymakers must adopt a holistic approach that considers the spatial, economic, and social dimensions of food availability to create urban environments that support healthy eating behaviors.

3 Challenges in Ensuring Equitable Food Access

3.1 Identifying Food Deserts, Food Swamps, and Food Mirages

Food deserts are areas, often in urban environments, where residents have limited access to affordable and nutritious food [21]. These areas typically lack supermarkets or grocery stores that offer fresh fruits, vegetables, and other healthy food options. Instead, food deserts often have an abundance of fast-food outlets and convenience stores that sell processed foods high in sugar, fat, and salt. The lack of access to healthy food in these areas contributes to poor dietary habits and increases the risk of chronic diseases such as obesity, diabetes, and cardiovascular disease. Food swamps refer to areas where unhealthy food options outnumber healthy ones. Unlike food deserts, food swamps may have grocery stores and supermarkets, but these are overshadowed by a high density of fast-food restaurants, convenience stores, and other outlets selling unhealthy, calorie-dense foods.

The prevalence of unhealthy food options in food swamps makes it challenging for residents to maintain a healthy diet, leading to similar health issues as those found in food deserts. Food mirages occur in areas where healthy food options are available, but they are not affordable for the local population. This phenomenon often affects low-income neighborhoods where residents may see healthy foods in nearby stores but cannot purchase them due to high prices. Food mirages highlight the importance of both physical and economic access to nutritious food. Addressing food mirages requires not only increasing the availability of healthy food but also ensuring that these foods are affordable for all residents.

Food deserts, food swamps, and food mirages are critical concepts in understanding food availability challenges. Identifying these food availability challenges requires a combination of geographic information systems (GIS) and demographic analysis [22]. GIS mapping can reveal spatial patterns in food access, highlighting areas with limited availability of grocery stores and high concentrations of fast-food outlets. Demographic data such as income, education, and ethnicity help identify populations most affected by food deserts, food swamps, and food mirages. By overlaying geographic and demographic data, planners can pinpoint priority areas for intervention.

3.2 Barriers to Healthy Food Access in Urban Areas

Economic barriers are significant obstacles to healthy food access. In many urban areas, the cost of nutritious foods like fruits, vegetables, and lean proteins is higher than processed and fast foods [23]. This price disparity makes it challenging for low-income residents to afford a healthy diet. Additionally, grocery stores may be less likely to be located in low-income neighborhoods, further limiting availability.

Physical barriers also play a crucial role in food access [24]. Residents in food deserts often face long distances to the nearest grocery store, making it difficult to purchase fresh food regularly. Limited public transportation options exacerbate this problem, particularly for those without access to a car. Walking or biking to a grocery store may be unsafe or impractical in areas with poor infrastructure.

3.3 Urban Planning and Policy Solutions

Zoning and land use regulations can significantly impact food availability [25]. Policies that encourage the development of grocery stores and farmers' markets in underserved areas can improve access to healthy foods. Conversely, zoning laws that limit the density of fast-food restaurants can help reduce the prevalence of unhealthy food options. Incentivizing mixed-use developments that incorporate food retail into residential areas can also promote healthier food environments.

Financial incentives and subsidies can encourage healthy food retailers to locate in underserved areas. Programs like tax breaks, grants, and low-interest loans can offset the higher costs and risks associated with opening grocery stores in low-income neighborhoods. Supporting local food producers through subsidies can also increase the availability of fresh, affordable produce.

4 Case Study: Changsha City, China

4.1 Different Concerns on Food Availability in Developed and Developing Countries

In developed nations, food availability issues often revolve around disparities in access to healthy food options within urban environments. For instance, in the United States, significant research has documented the prevalence of food deserts in low-income neighborhoods. These areas lack supermarkets and fresh food markets, instead featuring a higher density of fast food outlets and convenience stores. A notable example is the city of Chicago, where policy interventions, such as the opening of new grocery stores in underserved areas, have shown promise in addressing these disparities. Similarly, in the United Kingdom, the Healthy Start program provides vouchers for fresh fruits and vegetables to low-income families, demonstrating a targeted approach to improving food access.

In developing countries, food availability challenges often stem from broader systemic issues such as infrastructure deficiencies, economic constraints, and urbanization pressures. In India, rapid urbanization has led to the proliferation of informal food markets that may not always provide healthy options. Programs like the Public Distribution System (PDS) aim to ensure food security by distributing staple foods at subsidized rates, but there are ongoing efforts to diversify these offerings to include more nutritious options. In Kenya, initiatives like urban agriculture in Nairobi have been pivotal in improving food access in slum areas, where traditional retail options are scarce. These community-led projects not only enhance food security but also provide economic opportunities for residents.

4.2 Background and Context of Changsha City, China

Changsha, the capital of Hunan Province, is a rapidly growing urban center in central China. With a population of over 8 million, Changsha has experienced significant urbanization and economic development in recent decades. The city's expansion has led to diverse demographic changes, including an influx of rural migrants, increasing urban density, and evolving socio-economic structures. Understanding these trends is crucial for analyzing how urban development impacts food availability and public health. As Changsha continues to grow, ensuring that all residents have equitable access to nutritious food becomes increasingly important (Fig. 3).

The public health landscape of Changsha mirrors many of the challenges faced by other rapidly urbanizing cities in China. Non-communicable diseases (NCDs) such as cardiovascular diseases (CVD), diabetes, and obesity are on the rise, driven by changes in lifestyle, dietary habits, and environmental factors. The city has made strides in improving healthcare infrastructure and services, but disparities in health outcomes persist, particularly among low-income and marginalized populations. The prevalence of NCDs highlights the need for a comprehensive approach to urban health that includes improving neighborhood food environments. By addressing food

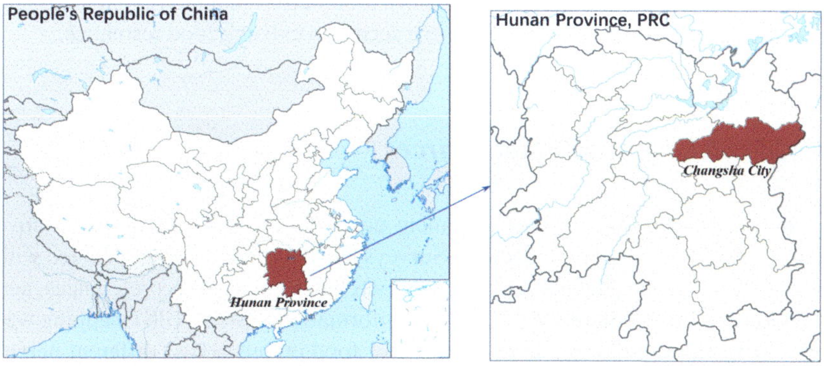

(a) Geographical Context of China: Highlighting Hunan Province

(b) Geographical Context of Hunan Province: Highlighting Changsha City

Fig. 3 Changsha City, China, as the study case

availability, Changsha can take significant steps toward reducing the burden of NCDs and enhancing overall public health.

4.3 Overview of Food Availability in Changsha's Neighborhoods

Changsha's food landscape comprises a variety of food outlets, including supermarkets, wet markets, convenience stores, and fast-food restaurants. Supermarkets and wet markets are primary sources of fresh produce, meat, and other essential groceries. Wet markets, in particular, are a traditional and popular choice among residents for purchasing fresh food items. However, the distribution of these outlets is uneven, with more affluent neighborhoods enjoying better access to a variety of healthy food options compared to lower-income areas. This disparity creates challenges for residents in underserved neighborhoods who may rely on convenience stores and fast-food outlets, which typically offer fewer healthy options. Mapping the distribution of these food outlets can provide insights into the areas that require targeted interventions to improve food access.

Socio-economic disparities significantly affect food availability in Changsha. Wealthier neighborhoods generally have more supermarkets and higher-quality food outlets, offering a wide range of nutritious foods. In contrast, lower-income areas often lack these resources, leading to a reliance on less healthy, processed foods. These disparities are compounded by factors such as education levels, employment status, and transportation access, which influence residents' ability to obtain healthy foods. Addressing these socio-economic disparities is crucial for promoting equitable access to nutritious food across all neighborhoods in Changsha. Strategies to bridge these gaps include subsidizing healthy food outlets in underserved areas and enhancing public transportation to improve access to existing food resources.

4.4 Methodology and Data Sources

The study on Changsha's food environment utilized a combination of quantitative and qualitative data collection methods. Surveys and interviews were conducted with residents to gather information on their food-purchasing behaviors, preferences, and perceptions of food availability. Geographic Information System (GIS) mapping was employed to analyze the spatial distribution of food outlets across different neighborhoods. Additionally, secondary data from government reports, health surveys, and academic studies provided a comprehensive understanding of the demographic and health profiles of Changsha's residents. This multi-method approach ensures a holistic analysis of the food environment and its impact on public health.

Several analytical approaches were applied to examine the relationship between neighborhood food availability and health outcomes in Changsha. Spatial analysis using GIS helped identify areas with high and low food availability, allowing for a visual representation of food deserts and swamps. Statistical analysis, including regression models, was used to explore associations between food outlet density, socio-economic factors, and health indicators such as obesity and CVD prevalence. Qualitative data from resident interviews provided contextual insights into the challenges and barriers faced by different communities. By integrating these analytical methods, the study offers a nuanced understanding of how food environments affect public health in Changsha.

5 Findings from Changsha City

5.1 Association Between Neighborhood Food Availability and Housing Prices

5.1.1 Pre and Post COVID-19 Analysis

The study of Changsha's neighborhood food environments revealed significant associations between food availability and housing prices, both before and after the COVID-19 pandemic. Before the pandemic, neighborhoods with higher housing prices tended to have better access to "unhealthy" late-night snack shops. Neighborhoods with higher housing prices did not have better access to "healthy" restaurants or food stores than those with lower housing prices before the pandemic. This trend indicates that high-income neighborhoods don't necessarily have more healthy food resources than low-income neighborhoods.

The COVID-19 pandemic further accentuated the health challenge for high-price developments. After the first round of the pandemic, the neighborhood availability of all types of food outlets decreased, except for late-night snack shops. Neighborhoods that saw a relative increase in housing prices suffered from a decrease in the availability of sit-down restaurants. Low housing prices might promote the availability of "healthy" restaurants and thereby partly protect residents from high-calorie and high-fat diets. The pandemic has thus underscored the critical role of housing market dynamics in affecting urban resilience of food availability.

5.1.2 Spatial and Temporal Trends

Spatial analysis of Changsha revealed distinct patterns in the distribution of food outlets and their correlation with housing prices. The core areas in the central city, occupied mainly by old residential developments and characterized by higher population density and better infrastructure, generally exhibited higher food outlet density and lower housing prices than other areas in the central city occupied by relatively new residential developments. Over time, the spatial disparities in food availability have persisted, with some neighborhoods experiencing improvements while others continue to face challenges. Although both the established and new neighborhoods experienced an increase or decrease in the number of various food stores before and after the Pandemic, the availability of food outlets in the established neighborhoods was always higher than that in the new ones.

Temporal trends indicate that efforts to keep a relatively low housing price are critical to keeping a high level of healthy food availability. We did not observe such temporal trends among new residential developments. Keeping housing prices low does not necessarily help access to healthy food in all neighborhoods, reflecting the complexity of the task of improving access to food. Addressing these spatial and temporal trends requires targeted interventions that focus on underserved neighborhoods to ensure equitable access to nutritious food for all residents.

5.2 Elderly Residents' Perception and Use of Daily Consumer Facilities

5.2.1 Survey Results and Analysis

The study surveyed elderly residents in various neighborhoods of Changsha to understand their perceptions and use of daily consumer facilities, a.k.a. food outlets such as supermarkets, restaurants, convenience stores, etc. The findings revealed that the elderly prioritize proximity and convenience when choosing where to shop for food. Many elderly respondents reported challenges in accessing food outlets due to physical mobility limitations, lack of transportation options, and the distance of facilities from their homes.

Analysis of the survey data showed that neighborhoods with higher densities of food outlets had higher satisfaction levels among elderly residents regarding food access. In contrast, areas with fewer food outlets reported lower satisfaction and higher levels of food insecurity among the elderly. The elderly's reliance on nearby food outlets underscores the importance of ensuring that these facilities are accessible and adequately stocked with healthy food options.

5.2.2 Impact of Distance and Facility Size

The distance between home and food outlets emerged as a significant factor influencing the elderly's food purchasing behaviors. Longer distances were associated with decreased frequency of visits to food outlets and higher reliance on less nutritious, easily accessible food options from convenience stores. Smaller, neighborhood-level stores, often found in closer proximity to residential areas, were frequented more regularly by the elderly compared to larger supermarkets that were farther away.

Facility size also played a role in food access for the elderly. Larger supermarkets, while offering a wider variety of products, were less accessible to elderly residents due to their locations and the time required to navigate these larger spaces. Conversely, smaller neighborhood stores, despite having limited product ranges, were more frequently visited due to their convenience. This finding highlights the need for a balanced approach in urban planning that integrates both large and small food outlets within close proximity to residential areas, particularly those with high elderly populations.

5.3 *Impact of Commuting Distances on Food Availability and Cardiovascular Health*

5.3.1 Moderating Effects of Commuting

The study explored the mediation effects of commuting distances on the relationship between neighborhood food availability and cardiovascular health among residents. It was found that longer commuting distances negatively impacted residents' dietary behaviors and overall health. Residents with longer commutes were less likely to shop at stores offering healthy food options. Possibly this is due to time constraints and fatigue, leading to increased reliance on fast food and convenience stores around home. This shift in dietary habits contributed to poorer cardiovascular health outcomes, including higher rates of hypertension, obesity, and diabetes (Fig. 4).

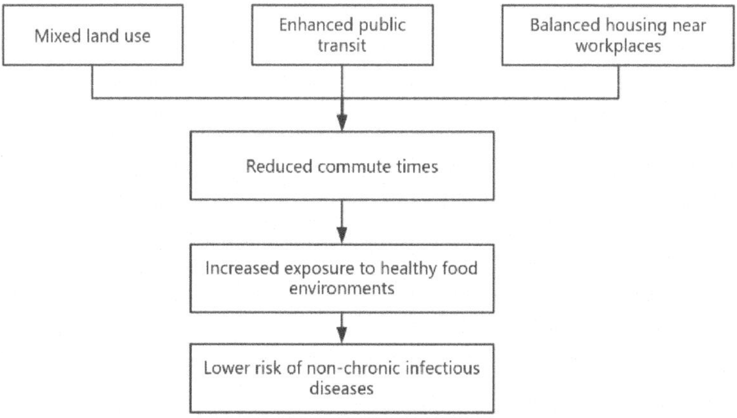

Fig. 4 Built environment factors affecting commuting time, exposure to healthy food environments, and risk of non-chronic infectious diseases. *Source* Drawn by Authors

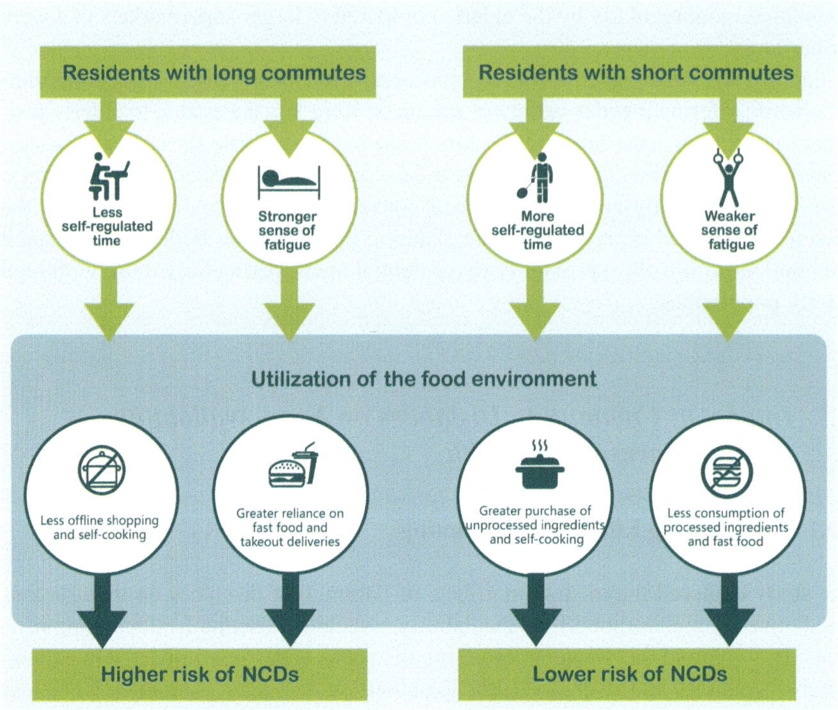

Fig. 5 Moderating effect of commuting on the relationship between food environment and NCDs. *Source* Drawn by Authors

The moderation analysis demonstrated that commuting distance serves as a significant barrier to accessing healthy food, thereby influencing health outcomes. This finding underscores the importance of considering commuting patterns in urban planning and public health strategies. Reducing commuting distances by promoting mixed-use development and improving public transportation can enhance food access and contribute to better health outcomes (Fig. 5).

5.3.2 Socio-Economic Moderators

The impact of commuting distances on food environment and cardiovascular health varied across different socio-economic groups. Lower-income residents were more adversely affected by long commutes, as they had fewer resources to mitigate the impact of limited food access. For instance, they were less likely to own private vehicles and more dependent on public transportation, which further restricted their access to distant food outlets.

Conversely, higher-income residents, despite facing long commutes, were better able to maintain healthier diets due to their greater access to resources such as private transportation and higher disposable incomes, allowing them to purchase healthier food options even from distant stores. This socio-economic disparity highlights the need for targeted interventions that address the unique challenges faced by lower-income populations in accessing healthy food. Ensuring equitable food access requires policies that reduce commuting burdens and enhance the availability of healthy food options within all neighborhoods, regardless of residents' socio-economic status.

6 Discussion: Implications for Urban Planning and Policy

6.1 Interpretation of Findings from Changsha: Key Insights and Surprising Results

The study of neighborhood food environments in Changsha revealed several key insights and unexpected results. Firstly, the significant correlation between food availability and housing prices underscored the value of low housing prices in keeping a healthy food environment around people's homes. This trend persisted and even intensified during the COVID-19 pandemic, highlighting the crucial role of housing prices in maintaining urban resilience. The pandemic served as a catalyst, amplifying the importance of having food outlets within close proximity to residential areas. This finding is consistent with global trends that emphasize the importance of food security in times of crisis.

Furthermore, the survey of elderly residents highlighted the specific challenges faced by this demographic in accessing healthy food. The reliance on nearby food outlets due to mobility issues and transportation barriers was a prominent theme. The elderly's preference for smaller, neighborhood stores over larger supermarkets due to convenience and accessibility emphasizes the need for urban planning that caters to the unique needs of different population groups.

The analysis of commuting distances and their impact on dietary behaviors and cardiovascular health provided additional insights. Long commuting distances were found to negatively impact health outcomes, leading to increased reliance on "unhealthy" fast food and late-night snack shops. This, in turn, contributed to poorer cardiovascular health outcomes. The moderating effect of commuting distance highlights the interconnectedness of urban planning, transportation, and public health.

6.2 Comparison with Global Trends

The findings from Changsha align with global trends observed in other urban contexts. Studies from various countries have demonstrated similar relationships between neighborhood food environments and health outcomes. For example, research in the United States and Europe has consistently shown that areas with better access to supermarkets and fresh food outlets have lower rates of obesity and chronic diseases compared to food deserts. The emphasis on local food availability during the COVID-19 pandemic is also a global phenomenon, with cities worldwide recognizing the importance of food security in enhancing urban resilience.

However, there are also unique aspects to the Changsha case that provide valuable insights. The specific challenges faced by the elderly and the significant impact of commuting distances on food access and health outcomes highlight the need for tailored interventions in urban planning. These findings underscore the importance of considering local context and population-specific needs when designing food environments.

6.3 *Recommendations for Improving Food Availability in Urban Settings*

6.3.1 Planning and Design Strategies

To improve food availability in urban settings, several planning and design strategies can be implemented. Firstly, mixed-use development that integrates residential areas with commercial spaces, including food outlets, can enhance access to nutritious food. Ensuring that food outlets are within walking distance of homes can reduce reliance on transportation and encourage healthier dietary behaviors. Urban planners should prioritize the inclusion of both supermarkets and smaller neighborhood stores in residential areas to cater to the diverse needs of the population.

Additionally, the design of public spaces and transportation infrastructure should facilitate easy access to food outlets. Enhancing pedestrian and cycling infrastructure can promote active transportation to food stores, contributing to both improved food access and increased physical activity. Public transportation routes should be planned to connect residential areas with key food retail locations, ensuring that all residents, including those without private vehicles, can access healthy food options.

6.3.2 Policy Interventions

Policy interventions play a crucial role in enhancing food availability in urban areas. Zoning regulations advocating reserving space suitable for the operation of supermarkets/wet markets can be used to incentivize the establishment of healthy food outlets in underserved areas. Implementing policies that restrict the density of fast-food outlets while promoting the establishment of outlets selling fresh produce can help create a healthier food environment.

Moreover, policies aimed at improving food affordability are essential. Subsidizing the cost of fresh fruits and vegetables or providing financial incentives for low-income families to purchase healthy food can address economic barriers to food access. Collaboration between government agencies, community organizations, and private sector stakeholders is necessary to develop and implement effective policies that promote food security and public health.

6.4 Strategies for Promoting Healthier and More Resilient Urban Environments

6.4.1 Community Engagement and Participation

Community engagement is vital for promoting healthier and more resilient urban environments. Involving residents in the planning and decision-making processes ensures that their needs and preferences are taken into account. Community gardens, farmers' markets, and food cooperatives are examples of initiatives that can enhance local food availability while fostering community cohesion. Encouraging local food production and supporting urban agriculture can also reduce food miles and promote environmental sustainability.

Educational programs that raise awareness about the importance of healthy eating and how to access nutritious food are also crucial. Partnering with local schools, community centers, and health organizations to deliver these programs can empower residents to make healthier food choices. Engaging the community in initiatives such as food audits and participatory mapping can provide valuable insights into local food environments and inform targeted interventions.

6.4.2 Integrated Approaches

Promoting healthier urban environments requires integrated approaches that consider the interplay between various factors influencing food access and public health. Collaborating with urban planners, public health officials, transportation authorities, and community organizations can ensure that strategies are comprehensive and effective. Integrated approaches should address the social, economic, and environmental determinants of health to create a holistic framework for improving food availability and urban resilience.

For instance, integrating health impact assessments into urban planning processes can help identify potential health implications of proposed developments and inform design decisions. Cross-sector partnerships can facilitate the sharing of resources and expertise, enabling the implementation of multifaceted interventions that address food access, transportation, housing, and public health concurrently.

7 Conclusion: Toward Healthier Urban Environments

7.1 Summary of Key Points

This chapter has explored the crucial role of neighborhood food availability in fostering healthier and safer urban environments, using Changsha, China as a case study. The findings underscore the significant impact of food availability on public health, particularly in the context of chronic diseases such as cardiovascular diseases (CVD), diabetes, and obesity. The study highlighted the correlation between housing prices and food outlet availability, emphasizing how economic factors influence access to nutritious food. It also revealed the specific challenges faced by elderly residents in accessing healthy food due to mobility issues and transportation barriers. Furthermore, the impact of commuting distances on dietary behaviors and cardiovascular health outcomes was a notable finding, illustrating the interconnectedness of urban planning, transportation, and public health.

Food availability is a fundamental determinant of health and well-being in urban environments. The presence of accessible, affordable, and nutritious food options directly influences dietary behaviors and health outcomes. Ensuring that urban populations have access to a variety of healthy foods can lead to better health outcomes, reduce the prevalence of chronic diseases, and alleviate the burden on healthcare systems. The COVID-19 pandemic has further highlighted the importance of food security and the need for resilient food systems that can withstand crises. Addressing food availability is not only a health imperative but also a vital component of creating vibrant, inclusive, and sustainable urban communities.

7.2 Future Directions for Research and Practice

While significant progress has been made in understanding the relationship between food availability and urban health, several gaps remain in the literature. There is a need for more longitudinal studies that examine the long-term effects of food environments on health outcomes. The current body of research is heavily reliant on cross-sectional studies or longitudinal studies within a narrow period, which provide valuable insights but are limited in their ability to establish causality. Additionally, there is a lack of studies that explore the impact of food environments on mental health and social well-being, areas that are increasingly recognized as important components of overall health. Further research is also needed to understand the unique challenges and needs of different population groups, such as the elderly, low-income families, and individuals with disabilities.

Emerging areas of interest in food availability research include the role of technology and digital platforms in enhancing food access. The rise of online grocery shopping and food delivery services presents new opportunities and challenges for ensuring equitable food access. Investigating how these technologies can be leveraged to improve food availability, particularly in underserved areas, is a promising area of future research. Additionally, there is growing interest in the concept of food sovereignty, which emphasizes the rights of communities to control their food systems. Exploring how urban planning can support food sovereignty initiatives and promote local food production is another important area for future research.

7.3 Call to Action for Urban Planners, Policymakers, and Community Stakeholders

7.3.1 Collaborative Efforts

Addressing the complex issue of food availability in urban environments requires collaborative efforts from multiple stakeholders. Urban planners, policymakers, public health officials, community organizations, and residents all have a role to play in creating healthier food environments. It is essential to foster partnerships and encourage active participation from community members in the planning and decision-making processes. Community engagement initiatives, such as participatory mapping and food audits, can provide valuable insights into local food environments and inform targeted interventions. By working together, stakeholders can develop comprehensive strategies that address the social, economic, and environmental determinants of food availability.

7.3.2 A Long-Term Vision

Creating healthier urban environments through improved food availability requires a long-term vision and sustained commitment. Policymakers should prioritize the integration of food availability into urban planning and development policies. This includes implementing zoning regulations that promote the establishment of food outlets in underserved areas, providing financial incentives for grocery stores to open in food deserts, and supporting local food production initiatives. Additionally, addressing the economic barriers to healthy food access, such as affordability and transportation costs, is crucial. Subsidizing the cost of fresh produce and providing financial incentives for low-income families to purchase healthy food can help ensure that all residents have access to nutritious food.

In conclusion, enhancing neighborhood food availability is essential for promoting healthier and more resilient urban environments. By prioritizing food availability in urban planning and public health efforts, cities can create vibrant, inclusive, and sustainable communities that support the health and well-being of all residents. The findings from this chapter provide valuable insights and practical recommendations for urban planners, policymakers, and community stakeholders. Through collaborative efforts and a long-term vision, it is possible to address the challenges of food availability and build healthier urban environments for the future.

References

1. Gao I, Kolak M (2019) Healthy access for healthy places: a multidimensional food access measure. arXiv preprint arXiv:1912
2. Walker RE, Keane CR, Burke JG (2010) Disparities and access to healthy food in the United States: a review of food deserts literature. Health & Place 16(5):876–884
3. Krieger J, Higgins DL (2002) Housing and health: time again for public health action. Am J Public Health 92(5):758–768
4. Maas J, Verheij RA, Groenewegen PP, de Vries S, Spreeuwenberg P (2006) Green space, urbanity, and health: how strong is the relation?. J Epidemiol Community Health 60(7):587–592
5. Morland K, Wing S, Diez Roux A (2002) The contextual effect of the local food environment on residents' diets: the atherosclerosis risk in communities study. Am J Public Health 92(11):1761–1767
6. Cummins S, Macintyre S (2006) Food environments and obesity—neighbourhood or nation? Int J Epidemiol 35(1):100–104
7. Walker RJ, Williams JS, Egede LE (2018) Influence of food insecurity on diabetes mellitus–a systematic review and meta-analysis. J Gen Intern Med 33(4):536–548
8. Gupta N, Khandelwal D (2019) Food insecurity: a risk factor for diabetes and cardiovascular disease in India. Indian Heart J 71(4):413–417
9. Fraser LK, Edwards KL, Cade J, Clarke GP (2010) The geography of fast food outlets: a review. Int J Environ Res Public Health 7(5):2290–2308
10. Li F, Harmer P, Cardinal BJ, Bosworth M, Johnson-Shelton D, Moore JM (2009) Built environment and 1-year change in weight and waist circumference in middle-aged and older adults: portland neighborhood environment and health study. Am J Epidemiol 169(4):401–408
11. Mackenbach JD, Rutter H, Compernolle S, Glonti K, Oppert JM, Charreire H, Brug J (2014) Obesogenic environments: a systematic review of the association between the physical environment and adult weight status, the SPOTLIGHT project. Obes Rev 15(5):453–463
12. Inoue M, Tsugane S (2009) Japan public health center-based prospective study group. Dietary fiber intake and risk of gastric cancer: the Japan public health center-based prospective study. Int J Cancer 125(9):2264–2271. https://doi.org/10.1016/j.arr.2017.06.002
13. Sadeghi N, Khosravi A (2021) Adherence to the Nordic diet is associated with lower depression, anxiety, and stress in individuals recovering from COVID-19. Nutr Neurosci 24(5):347–358. https://doi.org/10.1080/1028415X.2021.1915083
14. Loughrey D, Kelly M, Lawlor B (2017) The impact of the Mediterranean diet on the cognitive functioning of healthy older adults: a systematic review and meta-analysis. Ageing Res Rev 37:224–233. https://doi.org/10.1016/j.arr.2017.06.002
15. Dockery DW, Pope CA, Xu X, Spengler JD, Ware JH, Fay ME, Speizer FE (1993) An association between air pollution and mortality in six U.S. cities. N Engl J Med 329(24):1753–1759
16. Larson NI, Story MT, Nelson MC (2009) Neighborhood environments: disparities in access to healthy foods in the US. Am J Preventive Med 36(1):74–81
17. Cummins S, Macintyre S (2006) Food environments and obesity—neighbourhood or nation? Int J Epidemiol 35(1):100–104
18. Morland K, Wing S, Diez Roux A, Poole C (2002) Neighborhood characteristics associated with the location of food stores and food service places. Am J Preventive Med 22(1):23–29

19. Wang MC, Kim S, Gonzalez AA., MacLeod KE, Winkleby MA (2007) Socioeconomic and food-related physical characteristics of the neighborhood environment are associated with body mass index. J Epidemiol Community Health 61(6):491–498
20. Glanz K, Sallis JF, Saelens BE, Frank LD (2005) Healthy nutrition environments: concepts and measures. Am J Health Promot 19(5):330–333
21. Walker RE, Keane CR, Burke JG (2010) Disparities and access to healthy food in the United States: a review of food deserts literature. Health & Place 16(5):876–884
22. Rose D, Richards R (2004) Food store access and household fruit and vegetable use among participants in the US food stamp program. Public Health Nutr 7(8):1081–1088
23. Beaulac J, Kristjansson E, Cummins S (2009) A systematic review of food deserts, 1966–2007. Preventing Chronic Dis 6(3):A105
24. Larsen K, Gilliland J (2008) Mapping the evolution of 'food deserts' in a Canadian city: supermarket accessibility in London, Ontario, 1961–2005. Int J Health Geographics 7(1):16
25. Cummins S, Macintyre S (2006). Food environments and obesity—neighbourhood or nation? Int J Epidemiol 35(1):100–104

Low-Carbon Communities: Learning from Traditional Pedestrian-Friendly Neighbourhoods in Iran to Contribute to the Net-Zero Global Goal

Mousa Pazhuhan

Abstract Low-Carbon Communities (LCCs) play a crucial role in reducing greenhouse gas (GHG) emissions, improving sustainability, and enhancing urban livability. They also promote pedestrian-oriented living, reduce car dependence, and support a low-carbon lifestyle for residents. Iranian traditional neighborhoods are a prime example of LCCs due to their historical adaptation to the local climate and sustainable livelihood practices. This chapter aims to explore the architectural and urbanism principles practiced in the traditional neighborhoods of Iranian cities, thus learning some lessons for planning contemporary LCCs for domestic and abroad cities. At the same time, as the world works towards achieving a net zero carbon emissions target, it is important to learn from historical neighborhood architecture and design principles that have contributed to sustainable climate-friendly living throughout centuries. To achieve this, a qualitative archival method was employed, utilizing various publications to review the low-carbon characteristics of Iranian neighborhoods. The results revealed that traditional neighborhoods in old Iranian cities, Yazd, Shiraz, and Hamedan, exhibit critical principles that create car-free and pedestrian-friendly environments, including compact and mixed-use development, accessibility of daily services for all social classes, a human-scale physical structure, using vernacular material, integration of wind and solar energy Designing green buildings and provision of climatic comfort for inhabitants within the neighborhood and houses. By introducing the low-carbon urban design principles of traditional neighborhoods, this study provides insights for developing LCCs that can contribute to the global net-zero goal.

Keywords Low-carbon development · Pedestrian-friendly environment · Mitigation · Iranian traditional neighborhoods · Climate-friendly urban/architectural design

---

M. Pazhuhan (✉)
Geography and Urban Planning, Independent resilience researcher, Tehran, Iran
e-mail: pajoohan82@gmail.com

A. Cheshmehzangi et al. (eds.), *Designing Healthy Buildings and Communities*, Urban Sustainability, https://doi.org/10.1007/978-981-96-3057-8_7

1 Introduction

The global wide consequences of climate change are currently perceived as a ubiquitous threat, poised to catalyze significant global transformation. Given that cities serve as hubs for trade, industry, and growth, accounting for around 70% of total primary energy consumption, urban areas are now being recognized as a pivotal arena for addressing climate change [1, 2]. Accordingly, cities are key focal points for endeavors in adaptation and mitigation, and therefore, mitigation and adaptation strategies have increasingly been developed as main solutions to deal with changing climate [1]. Meanwhile, the urban level is increasingly recognized as a suitable scale for addressing climate change and promoting low-carbon urban development [3, 4, 2].

The Kyoto Protocol recognizes the concept of a "low carbon city" as a key element of sustainable urban development which involves not only reducing environmental emissions like carbon dioxide and advocating for eco-friendly products but also embracing a range of strategies to promote sustainability [5]. This includes the implementation of low-carbon projects across different levels of urban planning and development [5]. Indeed, urban neighborhood level, serving as the fundamental functional entity within a city, play a crucial role in decreasing carbon emissions and implementing the "low carbon city" strategy [6, 7]. Yu [8] asserts that efficient and low-carbon urban development necessitates the community as the fundamental unit. Many (e.g. [2, 6, 8, 9]) argue that the foundational attainment of low-carbon standards within the community is imperative to ensure the basic low-carbon principles of urban living and actualize city-wide low-carbon initiatives. Moreover, as Jank [10] concluded, communities have a crucial role in achieving a successful urban energy transition policy since it requires economically and technically viable solutions for optimized systems to manage the high costs of the energy transition and community involvement is crucial for effective implementation. Therefore, transitioning to an energy-efficient, low-carbon community stands as a fundamental policy imperative.

The emergence of LCCs and their significance in achieving low-carbon urban environments is a prominent subject in current academic discourse. The subject of carbon-neutral urbanization in Middle Eastern countries, such as Iran, with a rich history of thousands of years of architectural and urban development practices, has been notably neglected in international literature. Traditional neighborhood in the Iranian city represents an environmentally friendly urban life which is organized around some concepts of walkability, density, connectivity and accessibility, and mixed-use development [11]. These physical-spatial characteristics have been formed and practiced for years neighborhood residents and community planners alike in many Iranian cities [12]. This chapter explores the humane and environmentally conscious principles and values within architectural and neighborhood design practices in three traditional neighborhoods, Fahadan in Yazd, Sang-e-Siah in Shiraz, and Jowlan in Hamedan. It is noteworthy that all three case studies exist and are serving their residents. Moreover, this chapter aims to introduce some mitigation strategies applicable for contemporary practices for reaching LCCs by drawing lessons from those experiences have made Iranian traditional neighborhoods climate-friendly communities. Indeed, in the global effort to reach a net zero carbon emissions goal, it is crucial to draw insights from past urban planning and design approaches that have fostered enduring sustainability over many generations.

2 Low Carbon Community; Principles and Architectural and Urban Guidelines

The concept of "low carbon" was initially introduced in the UK energy white paper titled "Our Energy Future: Creating a Low Carbon Economy" in 2003, marking the beginning of research endeavors on this concept [3]. The term "low carbon community" denotes areas characterized by minimal carbon emissions, featuring eco-friendly surroundings, sustainable architecture, basic living practices, and efficient energy control [13]. As Cunha et al. [14] assert a low-carbon community acts as the basic building block of a low-carbon society and embodies the essence of a low-carbon urban area. It also can be seen as a cornerstone for laying the groundwork for a low-carbon society and achieving a low-carbon development strategy [13, 15]. The transition towards LCCs encompasses a diverse array of technologies, strategies, and low-carbon lifestyles [16–18]. This includes but is not limited to effective community planning, innovative green building technologies, adoption of renewable energy sources, sustainable transportation solutions, water recycling initiatives, waste reuse systems, and the shift of individual behaviors from high-energy and high-carbon consumption patterns to low-carbon lifestyles [17, 19].

Chen et al. [20], Xiao et al. [21], and Yin and Shi [22], have defined the concept of a low-carbon neighborhood as a tangible expression of collaboration and concerted efforts aimed at diminishing the carbon footprint associated with residents' lifestyle practices through the provision of conducive frameworks and operational mechanisms. According to Marchi and Gaspari [23] the concept of LCCs presents a novel framework for influencing the energy consumption behavior of end-users. Some others (e.g. [3, 8, 24, 25]) assert that LCCs are characterized by a compact spatial structure, sustainable transportation infrastructure, eco-friendly architectural design, effective energy utilization, waste material recycling and reuse practices, and community engagement. Indeed, Wang et al. [26], have focused on urban form and the physical and spatial aspect of LCCs and argue that the spatial configuration of the neighborhood, along with height density and construction pattern are crucial factors in energy consumption efficiency and the development of sustainable LCCs.

However, Goldstein et al. [27] and Huang et al. [28] have emphasized on the mixed land use and land regulations as main contributing factors to LCCs enhancement. From social and cultural perspective, many (e.g. [29–32]) have found that citizen participation and cultural factors are crucial in developing LCCs as incorporating citizen perspectives into the planning and implementation of LCCs guidelines will help the success rate of all plans and policies. Environmental factors particularly environmentally friendly livelihood [20, 33, 34], green infrastructures, water recycling, and consumption management [8, 33, 35], greenery configurations [36, 37] have also been identified as key elements in defining LCCs.

In addition, there are some guidelines developed in various countries to lead LCCs including the Building Research Establishment Environmental Assessment Method (BREEAM) for communities in the UK, the Leadership in Energy and Environmental Design for Neighborhood Development (LEED-ND) system in the US, and the Comprehensive Assessment System for Built Environment Efficiency (CASBEE) system in Japan, *Guideline for the Construction of Low carbon Pilot Communities in China,* guidelines for energy-efficient communities (Annex 51) in Germany [6, 10, 38]. According to Liu et al. [6], these guidelines mainly focus on six aspects of community, site selection, energy utilization, community water environment, transportation environment, ecological environment, and community planning and design in the early stage. Table 1 presents the most prevalent aspects, principles and foundations of LCCs in the literature.

As regards Table 1, various scholars have identified a wide spectrum of characteristics and indicators for LLCs that can be categorized in 6 general variables. These variables include land use management and transportation, urban form neighborhood facilities and infrastructures, neighborhood environment, community engagement and low carbon lifestyle. Historical neighborhoods in the case studies will be examined within the framework of LLCs principles and characteristics in the subsequent sections. This analysis aims to contrast the attributes that underscore Iranian neighborhoods as sustainable LLCs and characteristics recognized in existing literature as fundamental components of LLCs.

3 The Climatic Background of Iran

Iran is situated on a high plateau within latitudes ranging from 25 to 40 degrees north in an arid zone of the northern hemisphere of the Earth [43]. The arid deserts of Northern Africa and Saudi Arabia span from the Atlantic Ocean in Western Africa, through Iran, and culminate in Afghanistan and Turkmenistan [44].

Several Iranian environmental scholars have proposed distinct climatic classifications.

For instance, some researchers/scholars (e.g. [45] refer to UNEP's climate types for Iran including four types of "arid," "semi-arid," "humid," and "sub-humid" characterized by more than 98% of Iran.

Nevertheless, within the realms of architecture and urban planning, a consensus has emerged among the majority of academics on a four-fold climatic categorization [44].

1. Hot-dry climate (central plateau of Iran).
2. Mountainous cold climate (mountainous parts of western Iran).

Table 1 General variables and characteristics of LLCs

General variable	Characteristics	Justification
Land use management and transportation	Mixed land use, Land use compatibility, sustainable land use patterns, increasing density in neighborhood service points, incorporating environmental quality protection measures, Variety of local and small business activities, Affordable and high-quality transportation system, Providing pedestrian and bicycle access to the neighborhood services, Green Building	[3, 6, 8, 24, 25, 26]
Urban form	Compact spatial form, Reducing energy waste (insulation, absorption, etc.), Variety of housing types, Size and scale optimization, Climate-friendly Architectural design (light/ventilation/energy, etc.), Proper placement and orientation	[6, 27, 28, 26]
Neighborhood facilities and infrastructures	Management of energy conservation in the neighborhood, Maximum use of green and renewable energies (wind, radiation, geothermal, etc.), Minimal use of non-renewable energy, Wastewater collection, treatment and recovery system, Waste collection and waste recycling	[17, 19, 33, 35–37]
Neighborhood Environment	Management of water consumption and recycling, Conservation of soil resources, Preservation and promotion of vegetation and greenery in different forms	[8, 17, 19, 20, 33, 34]
Community engagement	The presence of different social groups of people in the neighborhood, Maximum participation in the realization of a low-carbon neighborhood, Citizens' trust in governmental and municipal initiatives for the implementation of a low-carbon community	[29–32]
Low carbon lifestyle	Emitting lowest carbon through walking and cycling, traveling by public transport, planting trees, using renewables (wind and solar for colling and heating) by architectural design, low carbon living	[20–23, 39–42]

3. Humid and moderate climate (southern borders of the Caspian Sea).
4. Hot-humid climate (northern borders of the Persian Gulf and the Oman Sea).

According to Memarian and Brown [46], Memarian [47], Memarian [48], Keshtkaran [44], de Andraca and Rodwell [49], the distinct characteristics of the diverse climatic regions in Iran have exerted notable influences on the architectural designs and selection of building materials in the urban areas of the country.

The inhabitants of the various regions of Iran have predominantly devised and implemented proficient solutions throughout centuries to establish architectural harmony with the local climate [50]. Amazingly, these solutions have effectively mitigated the adverse effects of the harsh climatic conditions and have also generated beneficial environmental conditions that are appreciated by the populace [44]. Generally, buildings and structures in this region have been thoughtfully integrated with the natural environment, leading traditional buildings in Iran to exhibit compatibility with and a harmonious relationship to natural conditions, unlike many modern constructions [51].

4 The Concept and Function of Neighbourhoods in the Iranian City

Neighborhoods (local communities) have been an integral part of the social, economic, managerial, and physical structure of traditional Iranian cities while they have had their own independent social and spatial organization simultaneously. The notion of neighborliness and the duty felt by residents towards one another are key characteristics of the historic neighborhoods in the Iranian city [52]. Another notable characteristic of traditional Iranian neighborhoods is their self-sufficiency, Apart from catering to the economic, social, and cultural requirements of the inhabitants, these neighborhoods exhibit a distinct administrative structure that operates semi-autonomously from the municipal authorities [53]. As Maroufi [54] asserts, Self-autonomy, self-sufficiency, and self-supportiveness constituted the primary social and administrative elements of neighborhoods in traditional Iranian urban centers.

Falamaki [55] delineates five characteristics concerning the principles of neighborhood life in Iranian traditional cities. The first characteristic pertains to the conditional and limited self-sufficiency and residential autonomy observed in each residential neighborhood within these communities. The second feature pertains to the presence of a distinct and constrained form of social life that operates independently from Bazaar influences, encompassing economic, political, and social mobility, within the urban layout and residential zones. The second feature contrasts the spatial complexity of the bazaar with the residential areas, emphasizing its role as a public space accessible to all citizens and newcomers to the city, characterized by a distinct residential ambiance. Fourth, the social behavior of residents in residential neighborhoods may exhibit distinctive characteristics influenced by the traditional traits of the families residing in these areas. And finally, the coexistence of families living in a neighborhood is affected by historical events and cultural identity [55].

In the Iranian city, form is intricately linked to meaning, where the physical structure derives its identity from the content, playing a crucial role in the social sustainability of the local community [56]. This significant correlation between form and content is shaped by the theoretical underpinnings of Iranian architectural and urban planning traditions, which, over millennia, have embedded the interconnection of form and content akin to the symbiosis of soul and body within their worldview [57]. Isfahan as one of the most important historical cities in Iran represents many of traditional characteristics of Iranian cities and its neighborhoods not many years ago were developed based on specific Iranian urbanism named Isfahan urbanism school [58]. The school exhibits a robust relationship with nature within the urban setting, where the central components of the neighborhoods, such as the square, significantly contribute to both physical and social structuring. Spatial balance and symmetry, the interplay of physical elements in fostering harmony, the influence of physical elements on the social identity of inhabitants, and the humane scale of the neighborhood environment represent additional physical attributes of this educational school (ibid.). Moreover, unity and plurality are other physical and spatial characteristics of traditional Iranian neighborhoods, based on which neighborhood spaces are organized in a specific area called neighborhood while accepting cultural and social diversity in their entirety [57]. Additionally, the predominant mode of transportation within the traditional Iranian neighborhood is by foot or bicycle, with minimal vehicular traffic circulating internally and passing traffic externally. Within this community, the neighborhood's framework is founded upon shared public interests, with inhabitants acknowledging a collective responsibility towards their residential neighborhood [59].

According to AbbasiHarofte and Sadeghian [60] and Hayati and Ameri [61] the traditional communities in Iranian cities have some climate-friendly characteristics in common including changing materials used for floors and walls (baseboards), covering spaces by putting roofs over them, creating split-level floors, differing proportions, ornamenting and changing light and shadow. These attributes were organized into three basic physical patterns including entrance, dead-end alleys, and sabat patterns which have been formed in the semi-public spaces of the neighborhoods. However, as AbbasiHarofte and Sadeghian [60] asserted while the communities in the historical cities of Iran commonly benefited from having neighborly interactions for a long time, this spatial organization along with their environmentally friendly physical patterns faded away from modern urban life and has caused serious social hazards.

5 Case Studies

This section will investigate three traditional neighborhoods in the old cities of Yazd, Shiraz, and Hamedan within an Iranian context to explore the principles and criteria of LCCs. The selection process for the case studies took into account diverse climatic conditions, including cold, arid, and hot environments. These neighborhoods are amongst the most vibrant and desirable residential communities in their own hometowns. However, the neighborhood values, such as climate-friendly environments, walkability, climatic adaptability, and energy efficiency, have been diminishing due to unbridled urbanization and the implementation of non-native and incompatible urban development plans that do not consider its physical and spatial context. Each case study consists of four subsections in which the city, the neighborhood, the

Architectural design principles and finally neighborhood design principles will be reviewed and analyzed.

5.1 Fahadaan Neighborhood, Yazd

The Historic City of Yazd was designated as a UNESCO World Heritage Site in 2017 [62]. Dating back to the fifth century, the city is situated in proximity to the Spice and Silk Roads, maintaining its significance as a prominent hub for textile production [49]. Yazd, *known as the city of windcatchers*, has played a decisive role and had a significant influence on the evolution and promotion of earthen architecture, technology, and urban planning in the Central and West Asian Region [63]. This is evident in terms of urban planning, material usage, and adaptation to its arid setting [63]. Moreover, the earthen architecture of Yazd exemplifies a climatically adapted urban layout, with the city's infrastructure centered on this construction material [46]. Prioritizing reduced consumption in both industry and housing is paramount, while harnessing solar radiation proves particularly advantageous in the desert setting [64]. Locals in the desert have devised clever solutions to combat the challenging environment, leading to the creation of impressive architectural and urban planning marvels [48]. Notable examples include qanats, badgirs, goudal-baqchehs, sabats, gozars, Abanbars, and various other architectural wonders [63].

The city is renowned as the *city of windcatchers*, structures designed to account for favorable and unfavorable winds, along with sandstorms, functioning proficiently as a conventional air-cooling mechanism. These windcatchers synergize with other cooling elements such as courtyards and fans [65]. In Yazd, climate change is fostering the emergence of initiatives aimed at establishing environmentally sustainable enterprises, cultivating ethical cultural tourism, and accentuating the ecological significance of urban areas [49].

5.1.1 The Neighborhood Background

The Fahadan neighborhood serves as the central nucleus of Yazd city, accommodating a population of approximately 12,000 individuals within an area spanning 66 hectares, representing close to 10% of the historical center of Yazd city [66]. The neighborhood exhibits enduring characteristics that have persisted to the present day. It comprises a cohesive collection of significant edifices, monuments, and numerous historical residences [67]. In Fahadan, the traditional society consists of long-term resident families who engage in community activities and religious events, fostering close relationships with neighbors and local figures [67].

5.1.2 Architectural Design Principles and Values

In Fahadan a particular architectural style is showcased, demonstrating the integration of traditional passive techniques to enhance living comfort and minimize technological dependence, adaptable to diverse climate conditions [49]. The vernacular form of the neighborhood's adobe houses has evolved over centuries in response to the city's desert climate. Adobe material is a strategic vernacular ingredient in the neighborhood architectural art as it is suitable for use in the hot and arid climate of Yazd owing to its significant capacity for thermal energy storage and regulation. Accordingly, the unique combination of environmental factors and the local abundance of high-quality soil for creating adobe units has established adobe architecture as the predominant construction style in the historic fabric of the Fahadan neighborhood [68].

Architecturally speaking, the houses are arranged around a central courtyard, organizing the residential spaces [69]. To mitigate the excessive heat of summer, particularly during the intense afternoon hours, the summer section, where families typically congregate, is oriented with its rear facing the Qibla (southward), known as "Nesar" in the local dialect [44]. Climate-friendly design considerations encompass sustainable utilization of natural materials, optimal utilization of available resources, and integration of wind and solar energy, defining key attributes of the eco-architectural approach in the neighborhood [44]. The wind-catcher stands as a prominent symbol within Yazd, especially within the Fahadan neighborhood. This element of vernacular architecture showcases remarkable architectural characteristics, distinctive structural attributes, and unparalleled efficacy in environmental regulation [69].

5.1.3 Neighborhood Design Principles and Values

According to Tavassoli [70] the architectural design of the Fahadan neighborhood aims to achieve high density and complexity in buildings to offer optimal weather protection during both winter and summer seasons while maximizing shading effects on residences and thoroughfares. Moreover, in line with the climate considerations streets and sidewalks are predominantly oriented in an east–west direction to shield inhabitants from excessive heat. Indeed, during sweltering summer periods, the narrow sidewalks, flanked by tall walls on either side, remain entirely shaded [70]. Fahadan neighborhood exemplifies a deliberate utilization of adobe and mud as the primary construction material. The utilization of adobe in local construction demonstrates practicality and feasibility in response to climatic conditions [44].

Moreover, houses are designed to face inner courtyards, minimizing openings to the streets while passages in the neighborhood are partially covered with "sabats," which are open-arched roofing structures used for shading and air circulation [65]. In addition, the wind tower (*Badgir*), known as an air trap or windcatcher, represents a remarkable innovation within the local architectural design of the Fahadan neighborhood, serving a climatic purpose. The windcatchers in the neighborhood, like to those found in other historic districts of the city, were designed to account for both favorable and the unfavorable winds, along with sandstorms [49]. *Badgirs* contributed significantly to regulating the indoor climate by harnessing the predominant winds and directing them through shafts into the underlying spaces. This mechanism led to achieving an internal air temperature that was notably lower by several degrees compared to the ambient temperature in shaded outdoor areas [71]. They functioned proficiently as a conventional air-cooling mechanism, synergizing with other cooling elements such as courtyards and fans [65]. Architects leverage these favorable winds during hot seasons through the utilization of an air trap mechanism [72]. Air traps primarily function to capture the favourable wind for cooling and ventilation purposes, channeling it towards key areas such as the main room, basement, and water storage facilities. Subsequently, the demand for industrial cooling systems powered by fossil fuels would be significantly reduced [44].

5.2 *Sang-E-Siah Neighborhood, Shiraz*

Situated in a basin on the plateau at an altitude of around 1200 m (4000 feet) above sea level, Shiraz functions as the capital of the historical region of Iran (Fars) [46]. Shiraz is the most populated city in southern Iran and the capital of Fars Province. The city has a mild climate with hot summers and mild winters (Mediterranean weather) with cold season rainfall [73]. Due to the height of the Shiraz plain and being located between two cold and hot climatic zones, its rainfall is more than in other parts of the province (ibid.).

5.2.1 The Neighbourhood Background

The Sang-e-Siah neighborhood's historical roots trace back to the Safari and Atabakan eras [74]. Notable cultural and historical landmarks in the area include the New Mosque, Bibi Dokhtaran Tomb, Siboyeh Mausoleum, Holy Mary Church, Siavashan Mosque, and numerous historically and architecturally significant residences [75]. The neighborhood in Shiraz is the oldest residential area in the city. Its high population density, diverse land use, reduced reliance on automobiles attributed to its walkability, and the compact and varied spatial structure have distinguished it as an exemplar of a low-carbon neighborhood within the Shiraz urban landscape [76]. This neighborhood is esteemed as a valuable public space due to its morphological characteristics and rich content, encompassing cultural and environmental significance [74].

The historical neighborhood of Sang-e-Siah is not only attracting a large number of people from inside and outside the country but also causing the economic and livelihood prosperity of the neighborhood and has reduced the deterioration process by encouraging other people to restore and renovate [76]. The neighborhood exhibits distinct functional zones, with a harmonious blend of residential and commercial activities prevailing at the peripheries, while the interior is characterized by a dominance of religious establishments followed by commercial enterprises (ibid.).

5.2.2 Architectural Design Principles and Values

As Memarian and Brown [46] assert from a spatial perspective, the impact of climate is prominently demonstrated by the arrangement of the house around the courtyard. Due to the high intensity of the summer sun and the occasional significant temperature variations, both on a seasonal and daily basis, occupants do not use all areas of the house simultaneously. So as to climate adaptation and consuming less energy for cooling, the architectural art in the Sang-e-Siah has been creative so that during the summer months and the warmer periods of spring and autumn, positioning the primary living quarters on the southern side of the courtyard is beneficial as it provides shade and vaulted chambers aid in air circulation [46]. In contrast, the northern section of the courtyard is optimal during the winter season as it receives warmth from the gentle winter sun and confined chambers aid in heat retention (ibid.). As a result, a comprehensive seasonal pattern of movement is established from the southern side of the courtyard during summer to the northern side during winter.

According to Memarian [47], secondary movements are noticeable throughout each season's daily progression. During the summer (starting from late May), individuals tend to occupy the southern rooms until the latter part of the morning, following which they might relocate briefly to the basement. During the winter season, the primary daily movement occurs within the northern winter quarters. In the evenings and night time, the compact, secluded chambers with limited entry points typically exhibit the highest temperatures, thus offering the greatest comfort [48].

For Memarian and Brown [46], in the architectural and design traditions of Shiraz neighbourhoods, climatic considerations were mostly reflected around the courtyard. Due to the high solar radiation during summer and the significant temperature fluctuations, residents do not utilize all areas of the house simultaneously. So as to reduce energy consumption during the summer months and the warmer periods of spring and autumn, positioning the primary living quarters on the southern side of the courtyard is a beneficial strategy as it ensures shade for optimal comfort [46]. The division of accommodation along the north-south axis is attributed to the orientation of the courtyard, primarily driven by the imperative to optimize seasonal utilization (ibid.).

5.2.3 Neighborhood Design Principles and Values

The traditional neighborhood form significantly influences household travel patterns and automobile reliance. Residents of Sange-Siah exhibit notably lower carbon emissions compared to other urban neighborhoods [77]. The study of Abbasi and Hajipour [77] showed that residents of the neighborhood exhibit reduced reliance on personal vehicles leading to lower emissions compared to inhabitants in surrounding areas.

The researchers have identified the compact spatial form of the neighborhood, walkability, mixed land uses, and well-distributed and accessible various land uses, as the underlying cause of this issue Abbasi and Hajipour [77].

According to Bonine [78], the noteworthy aspect of house orientation lies in its alignment with the adjacent streets; as the main streets alter their direction, the houses follow suit. Consequently, the alignment of residential dwellings demonstrates a significant degree of alignment with the urban layout in the neighborhood, suggesting that the arrangement of housing units is influenced by the configuration of the streets, rather than the opposite scenario. The climate, in conjunction with specific local topographic characteristics and the historical boundaries of pre-urban fields and orchards, plays a pivotal role in shaping the spatial configuration of the neighborhood ([46, 78].

The other factors that have had significant effect on the spatial form of the neighborhood are the slope the land and the direction of water channels. Moreover, the lines of the neighborhood streets drew their orientation from these two natural factors [48]. The central passage in the neighborhood, known as Sange Siah Passage, was once considered an important passage of the whole city that had formed important spaces and activities around it [79]. Currently, this passage plays a crucial role in the historical fabric of Shiraz, serving as a hub for daily necessities of the residents, in conjunction with a lively neighborhood center. It significantly contributes to diminishing car reliance and promoting pedestrian-friendly environments [80].

5.3 *Jowlan Neighborhood, Hamedan City*

Situated in western Iran, Hamedan served as the inaugural capital of the nation and stands as one of Iran's most ancient cities. The rich historical strata of Hamedan have engendered the development of significant architectural structures and environments across ages [81].

5.3.1 The Neighborhood Background

Jowlan neighborhood is situated within Municipality District 3 of Hamedan city, where a significant portion of the city's historic textures are concentrated. The neighborhood has preserved its scale and size across various epochs owing to its adjacency to the historic city of Hegmataneh (Hegmetane Hill) [81]. The neighborhood has followed a compact growth pattern as opposed to sprawled development, offers diverse housing options for various socioeconomic strata, and features a form and structure that have remained largely unaffected by adverse alterations and human interference due to the neighbourhood's natural conditions [82].

5.3.2 Architectural Design Principles and Values

In the architectural design of the Jowlan neighborhood, the integration of wind and solar energy sources to reduce energy consumption has been a key focus for traditional architects, aligning with the principles of sustainable architecture [83]. The city of Hamedan historically exhibits extreme cold in winter and mild weather in summer, along with intense sunlight, significant diurnal temperature variation, cold winter winds, and substantial snowfall. These climatic conditions have influenced the architectural features of traditional Hamedani houses, which typically incorporate elements such as a central courtyard, hall, Shahneshin, Sedari, Dalan, Iwan, Sizan, and Hozkhaneh [82].

In the historical context, the prevalent architectural layout in Hamedan featured a primary orientation along the north-south axis, whereby the southern facade of the edifice was strategically designed to capture increased sunlight during the winter season. The indigenous dwellings within the neighborhood typically comprised structures spanning two to three stories, with the ground level often situated approximately half to one meter below the surface [84].

One notable architectural feature of the neighborhood is the substantial quantity of rooms, occasionally numbering up to 40. This design element has been posited to facilitate familial togetherness, offer assistance to newlyweds, and encourage engagement in communal and social endeavors [81]. Many of the historic residences within the neighborhood featured subterranean passages known as "naqb" that interconnected multiple dwellings. Naqb is a type of home cooling system that has been widely used in the architecture of neighborhood houses. Naqb is, in fact, an underground channel that uses the earth's moisture to cool the air [85].

The utilization of high thermal capacity materials, such as straw for heat regulation within the building envelope, alongside the incorporation of stone, exemplifies the astute strategies employed by architects in the architectural design of residential dwellings. These climatic measures aim to optimize solar exposure during winter months while safeguarding against chilling winds [83].

The cold climatic conditions in Hamedan necessitated the construction of dense urban fabric and interconnected buildings to minimize the exposure of warm residential spaces to the harsh external environment. Consequently, the design of neighborhood houses featured smaller porches compared to houses in hot, arid regions like Yazd, primarily serving the purpose of shielding entrances and windows from snow buildup [85]. Moreover, the courtyards of the houses typically sat at a lower elevation than the sidewalks, while the room roofs were constructed to minimize heat retention at reduced heights (ibid.).

The architectural form of the houses, akin to the urban form, was meticulously planned and implemented in response to the regional climate, particularly addressing extreme cold conditions. The utilization of compact forms, such as rectangular prisms, with a minimized ratio of external surface area to internal volume, is well-suited for frigid environments [81]. To harness solar thermal energy, the exterior surfaces of the houses are covered in dark colors. The neighborhood houses are structured around a central courtyard, with interconnected spaces facilitating movement between areas while minimizing exposure to external surroundings [85].

5.3.3 Neighborhood Design Principles and Values

The neighborhood has undoubtedly had a dynamic atmosphere in the past and responsive to the social life of its residents. In the recent past, low-carbon and locally sourced materials predominated in the construction of residential and urban structures [85]. Jowlan is characterized by its close proximity to public transportation options and ample room for pedestrians and cyclists. The wide alleys designed with humans in mind have led to a reduction in vehicle traffic interacting with riders, consequently significantly lowering rider traffic in the area. Also, the compact and dense texture of the neighborhood has prevented the use of the neighborhood as a parking lot [82, 86]. The strategic positioning of amenities and the neighborhood's self-reliance in fulfilling residents' daily requirements which is organized around local squares have significantly minimized the daily travel demands of its inhabitants. Consequently, a majority of the residents fulfill their daily needs on foot, obviating the necessity for vehicular transportation. The human scale and identity of Jowlan neighborhood continue to attract its original residents due to its authenticity [85]. Moreover, the neighborhood boasts a diverse environment, encompassing well-maintained green areas and a variety of wildlife, notably birds [83, 85].

Small and enclosed urban spaces, dense urban fabric and connected buildings and main alleys and passages parallel to the ground level line and often with a small width are the main physical characteristics of the neighborhood, which are designed according to the climatic conditions [81]. Nevertheless, the neighborhood has recently experienced inappropriate and unsustainable physical interventions that could jeopardize its invaluable climate-friendly spatial and physical characteristics if such interventions persist [81].

6 Discussion and Implications

Today, despite the utilization of modern construction technologies and advancements in urban living within new residential neighborhoods, numerous climatic challenges persist in the city, predominantly stemming from excessive energy consumption. In contrast, traditional neighborhoods achieved optimal climatic comfort and spatial usability in homes and communities through sustainable climate-friendly architectural and urban planning approaches with minimal energy usage. Hence, it can be asserted that traditional neighbourhoods in Iranian cities embodied low-carbon and eco-friendly characteristics, establishing a symbiotic and beneficial connection with climatic factors. Moreover, architects and urban planners aligned these communities with the local climate, configuring houses, sidewalks, and infrastructure to optimize energy efficiency and safeguard inhabitants from extreme temperatures.

Traditional Iranian architecture and urbanism are largely influenced by factors such as climate, geography, available materials, and cultural beliefs. Traditional Iranian masons and builders developed techniques to improve architectural sustainability by utilizing natural materials, all without the aid of modern technologies [44]. Therefore, the Iranian neighborhoods are traditionally low carbon communities by default since nature-based materials were used for buildings and a climate friendly urbanism has been practiced. Moreover, regarding the literature LCCs have a strong tie with low carbon lifestyles and it is evident this tie is strongly working in the three investigated neighborhoods. Indeed, inhabitants of the three case studies have had low carbon lives since they traditionally use energy efficient approach in their daily routines by walking, using natural energy resources for cooling and heating and benefiting from climate friendly environment of the neighborhood.

In the historic neighborhoods of Iranian cities architecture has established a symbiotic and fruitful connection with the local culture, climate, and lifestyle of the residents. This synergy has led to enhanced thermal comfort, preservation of indigenous identity, and a tranquil domestic environment that prioritizes environmental and climatic considerations. It's important to note that these neighborhoods already exist in the structures of Iranian cities and are serving their residents alongside the new neighborhoods. It's interesting that all the research comparing old and new neighborhoods in terms of quality of life, social capital, low carbon indicators, etc. (e.g. [77, 87, 88]) shows that older neighborhoods perform much better than new neighborhoods in terms of environmental and climatic indicators.

The Chart 1 articulates some lessons that can be learned from architecture and urban planning and design principles practiced in the case studies.

7 Concluding Remarks

The results of the current study show that traditional architecture and urbanism practiced in the case studies were climate oriented and humane. Moreover, the findings of the current study articulate that some significant and deep relationships among people, climate, culture, lifestyle, architecture, urbanism and neighborhood environment have had a crucial role in socio-environmental sustainability and healthy community. The primary concern in the current urban planning and architecture of contemporary cities, which are heavily carbon-oriented, lies in the disconnect between traditional architectural styles and contemporary requirements. To address this challenge, one viable approach involves integrating historical architectural and urbanism practices with modern technological advancements to achieve a more cohesive and sustainable architectural framework.

The necessity to revisit historical solutions to accommodate severe weather conditions stems from the fact that past architectural designs evolved through a longstanding intergenerational transmission of local architectural practices, rigorously

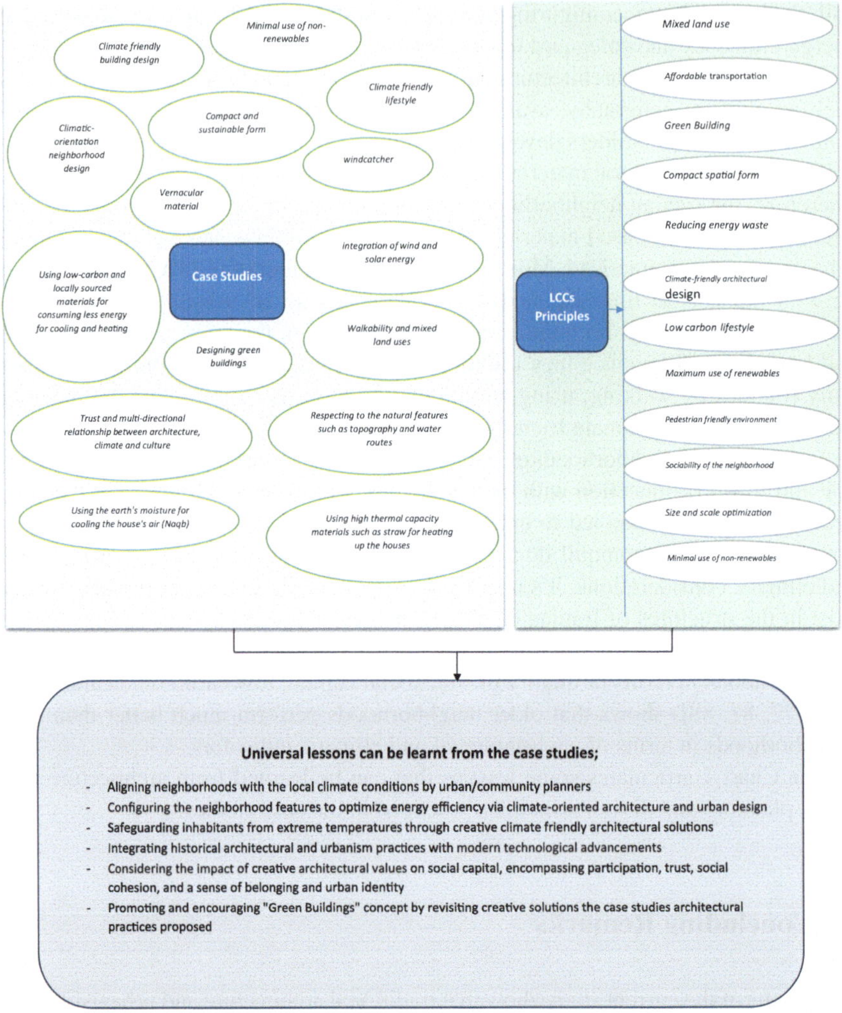

Chart 1 The most important lessons can be learnt from architectural and urbanism practices of the case studies for the contemporary LCCs

tested for stability and endurance through centuries of trial and error. In addition, the impact of architecture on social capital, encompassing participation, trust, social cohesion, and a sense of belonging and urban identity, as observed in the case studies, offers valuable insights for the transition of contemporary neighborhoods towards low-carbon models. This influence has been ingrained in the fabric of longstanding communities for generations.

What is more, the favorable and constructive impact has engendered trust in the architectural legacy of past generations among residents of these historic neighborhoods, leading to the construction of new buildings that adhere to established architectural principles and criteria. Consequently, a sustainable architectural heritage, compatible with the climate and low energy consumption, has been upheld across multiple generations within the community. In addition, the utilization of vernacular materials in traditional neighborhoods of Iranian cities has significantly reduced energy consumption and enhanced climate comfort for residents. These principles can be viewed as the foundation of "Green Buildings" a prominent topic in contemporary sustainable and climate-responsive architecture discussions.

All of the above-mentioned experiences about the case studies can be formulated in the form of lessons for climate planners and architects in all the countries with similar climatic conditions. As the results propose, LCCs would be achievable through if the connection between traditional architectural and urbanism styles and contemporary requirements are well redefined.

Acknowledgements The author acknowledges all the book editors who have kindly helped the quality of the chapter by their invaluable comments and guidance

References

1. Sharifi A (2021,January) Co-benefits and synergies between urban climate change mitigation and adaptation measures: a literature review: Sci Total Environ 750:141642
2. Zhao S, Feng T, Tie X, Long X, Li G, Cao J, An Z (2018) Impact of climate change on Siberian High and wintertime air pollution in China in the past two decades. Earth's Future 6(2):118–133
3. Wang X, Wang G, Chen T, Zeng Z, Heng CK (2023) Low-carbon city and its future research trends: a bibliometric analysis and systematic review. Sustain Cities Soc 90:104381
4. Pan A, Zhang W, Shi X, Dai L (2022) Climate policy and low-carbon innovation: evidence from low-carbon city pilots in China. Energy Econ 112:106129
5. Cheshmehzangi A, Xie L, Tan-Mullins M (2018) The role of international actors in low-carbon transitions of Shenzhen's International Low-Carbon City in China. Cities 74:64–74
6. Liu T, Wang Y, Li H, Qi Y (2021) China's low-carbon governance at community level: a case study in Min'an community, Beijing. J Clean Prod 311:127530
7. Roosta M, Javadpoor M, Ebadi M (2020) Developing a "low-carbon neighborhood model" in order to implement it in the urban planning & design. Urban Plann Knowl 4(1):33–48. https://doi.org/10.22124/upk.2020.15513.1383
8. Yu Y (2022) Explore the theoretical basis and implementation strategy of low-carbon urban community planning. Front Environ Sci 10:989318
9. Fang K, Azizan SA, Wu Y (2023) Low-carbon community regeneration in China: a case study in Dadong. Sustainability 15(5):4136
10. Jank R (2017) Annex 51: Case studies and guidelines for energy efficient communities. Energy Build 154:529–537
11. Jamei E, Ahmadi K, Chau HW, Seyedmahmoudian M, Horan B, Stojcevski A (2021) Urban design and walkability: lessons learnt from Iranian traditional cities. Sustainability 13(10):5731
12. Ebrahimpour-Masoumi H (2014) Neo-traditional neighbourhood design concepts based on the urban form of the historical cities of central Iran. Int J Sustain Soc 6(3):264–286

13. Li Y, Li S, Xia S, Li B, Zhang X, Wang B, Zheng W (2023) A review on the policy, technology and evaluation method of low-carbon buildings and communities. Energies 16(4):1773
14. Cunha FBF, Carani C, Castro NCA, C, Silva MS, Torres EA (2021) Transitioning to a low carbon society through energy communities: Lessons learned from Brazil and Italy. Energy Res Soc Sci 75:101994
15. Torabi Moghadam S, Di Nicoli MV, Manzo S, Lombardi P (2020) Mainstreaming energy communities in the transition to a low-carbon future: a methodological approach. Energies 13(7):1597
16. Giuzio GF, Forzano C, Barone G, Buonomano A (2024) Accelerating the low-carbon transition: technological advancements and challenges for the sustainable development of energy, water, and environment systems. Energy Rep 11:4676–4687
17. Griffiths S, Sovacool BK (2020) Rethinking the future low-carbon city: carbon neutrality, green design, and sustainability tensions in the making of Masdar City. Energy Res Soc Sci 62:101368
18. Jain AK (2023) Climate resilient, green and low carbon built environment. Springer Nature
19. Wu Y, Martens P, Krafft T (2022) Public awareness, lifestyle and low-carbon city transformation in China: a systematic literature review. Sustainability 14(16):10121
20. Chen L, Msigwa G, Yang M, Osman AI, Fawzy S, Rooney DW, Yap PS (2022) Strategies to achieve a carbon neutral society: a review. Environ Chem Lett 20(4):2277–2310
21. Xiao J, Zhen Z, Tian L, Su B, Chen H, Zhu AX (2021) Green behavior towards low-carbon society: theory, measurement and action. J Clean Prod 278:123765
22. Yin J, Shi S (2021) Social interaction and the formation of residents' low-carbon consumption behaviors: an embeddedness perspective. Resour Conserv Recycl 164:105116
23. Marchi L, Gaspari J (2023) Energy conservation at home: a critical review on the role of end-user behavior. Energies 16(22):7596
24. Sun L, Liu W, Li Z, Cai B, Fujii M, Luo X, Le Y (2021) Spatial and structural characteristics of CO_2 emissions in East Asian megacities and its indication for low-carbon city development. Appl Energy 284:116400
25. Falahatkar S, Rezaei F (2020) Towards low carbon cities: Spatio-temporal dynamics of urban form and carbon dioxide emissions. Remote Sens Appl Soc Environ 18:100317
26. Wang S, Wang J, Fang C, Li S (2019) Estimating the impacts of urban form on CO2 emission efficiency in the Pearl River Delta, China. Cities 85: 117–129
27. Goldstein B, Gounaridis D, Newell JP (2020) The carbon footprint of household energy use in the United States. Proc Natl Acad Sci 117(32):19122–19130
28. Huang Z, Yu H, Peng Z, Zhao M (2015) Methods and tools for community energy planning: a review. Renew Sustain Energy Rev 42:1335–1348
29. Lennon B, Dunphy NP, Sanvicente E (2019) Community acceptability and the energy transition: a citizens' perspective. Energy Sustain Soc 9(1):1–18
30. Sovacool BK, Griffiths S (2020) The cultural barriers to a low-carbon future: a review of six mobility and energy transitions across 28 countries. Renew Sustain Energy Rev 119:109569
31. Chilvers J, Pallett H, Hargreaves T (2018) Ecologies of participation in socio-technical change: the case of energy system transitions. Energy Res Soc Sci 42:199–210
32. Mundaca L, Busch H, Schwer S (2018) 'Successful low-carbon energy transitions at the community level? An energy justice perspective. Appl Energy 218:292–303
33. De Jong M, Joss S, Schraven D, Zhan C, Weijnen M (2015) Sustainable–smart–resilient–low carbon–eco–knowledge cities; making sense of a multitude of concepts promoting sustainable urbanization. J Clean Prod 109:25–38
34. Gui EM, MacGill I (2018) Typology of future clean energy communities: an exploratory structure, opportunities, and challenges. Energy Res Soc Sci 35:94–107
35. Parra D, Swierczynski M, Stroe DI, Norman SA, Abdon A, Worlitschek J, Patel MK (2017) An interdisciplinary review of energy storage for communities: challenges and perspectives. Renew Sustain Energy Rev 79:730–749
36. Xi C, Ding J, Wang J, Feng Z, Cao SJ (2022) Nature-based solution of greenery configuration design by comprehensive benefit evaluation of microclimate environment and carbon sequestration. Energy Build 270:112264

37. Privitera R, Palermo V, Martinico F, Fichera A, La Rosa D (2018) Towards lower carbon cities: Urban morphology contribution in climate change adaptation strategies. Eur Plan Stud 26(4):812–837
38. McGranahan G, Balk D, Anderson B (2007) The rising tide: assessing the risks of climate change and human settlements in low elevation coastal zones. Environ Urban 19(1):17–37
39. Pettifor H, Agnew M, Wilson C (2023) A framework for measuring and modelling low-carbon lifestyles. Glob Environ Chang 82:102739
40. Agnew MD, Pettifor H, Wilson C (2023) Lifestyle, an integrative concept: cross-disciplinary insights for low-carbon research. Wiley Interdisc Rev Energy Environ 12(6):e490
41. Schanes K, Giljum S, Hertwich E (2016) Low carbon lifestyles: A framework to structure consumption strategies and options to reduce carbon footprints. J Clean Prod 139:1033–1043
42. Büchs M, Saunders C, Wallbridge R, Smith G, Bardsley N (2015) Identifying and explaining framing strategies of low carbon lifestyle movement organizations. Glob Environ Chang 35:307–315
43. Roshan G, Grab SW, Najafi MS (2020) The role of physical geographic parameters affecting past (1980–2010) and future (2020–2049) thermal stress in Iran. Nat Hazards 102:365–399
44. Keshtkaran P (2011) Harmonization between climate and architecture in vernacular heritage: a case study in Yazd, Iran 2011 International Conference on Green Buildings and Sustainable Cities, Tehran
45. Raziei T (2022) Climate of Iran according to Köppen-Geiger, Feddema, and UNEP climate classifications. Theor Appl Climatol 148:1395–1416. https://doi.org/10.1007/s00704-022-03992-y
46. Memarian G, Brown FE (2003) Climate, culture and region, aspects of the traditional courtyard house in Iran. J Archit Plann Res 20(3)
47. Memarian G (1993) Ashnaii ba Memari Maskooni Irani: Ghuneh Shenasi Darunghara(An introduction to house typology in Iran: courtyard houses). Iran University of Science and Technology, Tehran
48. Memarian, G (1998) House typology in Iran with special reference to shiraz, Unpublished doctoral dissertation, University of Manchester, Manchester, UK
49. de Andraca IL, Rodwell D (2021) Environmental value for heritage conservation and urban sustainability: adaptation in widely divergent climatic conditions. Hist Environ Policy Pract 12(3–4):478–512. https://doi.org/10.1080/17567505.2021.1991104
50. Habibi K, Hoseini SM, Dehshti M, Khanian M, Mosavi A (2020) The impact of natural elements on environmental comfort in the Iranian-Islamic historical city of Isfahan. Int J Environ Res Public Health 17(16):5776
51. Kamali M (2024) The interaction of history and modern thought in the creation of Iran's architecture by investigating the approaches of past-oriented architecture. Front Architect Res.
52. Bahrimoqaddam A, Yousefifar S (2013) Social-cultural factors in forming district in Iranian-Islamic City and its functions. J Islamic Hist Civilisation 9(17):101–116
53. Kheirabadi M (1997) Iranian cities: formation and development. University of Texas Press, Austin
54. Maroufi H (2018) Restudying the Iranian (Islamic) city analyzing the institutions of urban governmentality in Iranian cities and their influences on urban space during Qajar and Safavid Era. J Iran Architect Urbanism 9(15):33–45
55. Falamaki S (2005) Simultaneous localisation and mapping on a model off-road vehicle (Doctoral dissertation, The University of New South Wales)
56. Sharifi A, Murayama A (2013) Changes in the traditional urban form and the social sustainability of contemporary cities: a case study of Iranian cities. Habitat Int 38:126–134
57. Pazhuhan M, Zayyari K, Ghasemzadeh B, Qurbani H (2015) Urban identity and Iranian new towns. J Urban Reg Anal 7(1):83–100
58. Habibi M (2005) Az Shar Ta Shahr, Tahlili Tarikhi az mafhum-i shahr va simay-i kalbodi-i an tafakor-o Tassor
59. Einifar A (2007) The role of primitive patterns in the contemporary residential design. Tehran Honar-ha-ye Ziba, 32(32)

60. Abbasi Harofteh M, Sadeghian A (2020) Spatial Patterns of Neighborhoods in the Historic City of Yazd; Determinants, Architectural Solutions and Principles of Neighborhood Architecture. Monthly Sci J Bagh-e Nazar 16(80):5–16
61. Hayati H, Ameri MA (2023) Physical patterns and neighborhood architecture in the historical context of Dezful City (principles of neighborhood architecture and architectural tools). Armanshahr Archit Urban Dev 16(44):29–47. https://doi.org/10.22034/aaud.2023.354887.2697
62. Nasrolahi A, Roux JM, Ghasvarian Jahromi L, Khalili M (2019) Assessment of local people opinion after world heritage site designation, case study: historic city of Yazd, Iran. Heritage 2(2):1739–1747
63. UNESCO (2018) Historic city of Yazd. Available online: https://whc.unesco.org/en/list/1544/documents (accessed on 10 June 2018).
64. Gharehgolchian A (2010) Concept of solar sustainable heat and cool. The first national conference of contemporary architecture & urbanization. Shiraz, Iran.
65. Moradivand S (2021) Investigating the effects of modernity on the changes in the spatial-physical structure of Yazd WHC. J Geogr New Stud Archit Urbanism.
66. Hekmatnia H (2020) Investigating role of public participation in the rehabilitation old texture, Fahadan neighborhood in Yazd. Human Geogr Res 52(2):639–656
67. Fallah Aliabadi S, Sarsangi A, Modiri E (2015) The social and physical vulnerability assessment of old texture against earthquake (case study: Fahadan district in Yazd City). Arab J Geosci 8:10775–10787
68. Aqasafari A, Amiri J, Beheshtian DMJ (2011) The quantitative and qualitative characteristics of housing in the historical texture of the Yazd city (1976–2006). J Stud Iran Islamic City 2:67–82
69. Azizi-Bondarabadi H (2022) Local seismic culture in Iranian vernacular architecture: evidence from Yazd earthen architecture. Built Heritage 6:5. https://doi.org/10.1186/s43238-022-00051-3
70. Tavassoli M (2002) Urban structure and architecture in the hot arid zone of Iran. Payam and Peivand Publication, Tehran
71. Sangdeh PK, Nasrollahi N (2022) Windcatchers and their applications in contemporary architecture. Energy Built Environ 3(1):56–72
72. Mahmoodi M (2009) The air trap. Yazda Publication, Tehran
88 Pazhuhan M, Amirzadeh M (2024) Community flood resilience assessment of Saadi neighborhood, Shiraz, Iran. J Risk Res 27(1):21–45
74. Rabiee R, Falamaki MM (2022) Analysis of the roots of identity crisis and its impact on the quality level of residential areas (Case study:Sang-e siah neighborhood of Shiraz)
75. Hamidi MA, Khakzand M, Faizi M (2021) Investigation on attaining healthy housing by increasing sense of coherence in historical context with emphasize on salutogenic approach: Sang-e-Siyah community, Shiraz, Iran. J Iran Archit Urbanism 12(2):137–151
76. Sadeghi AR, Ebadi M, Shams F, Jangjoo S (2022) Human-built environment interactions: the relationship between subjective well-being and perceived neighborhood environment characteristics. Sci Rep 12(1):21844
77. Abbasi H, Hajipour K (2014) An empirical analysis of the influence of urban form on travel behavior in different urban neighborhoods of Shiraz. The Monthly Sci J Bagh-e Nazar 11(29):23–32
78. Bonine ME (1979) The morphogenesis of Iranian cities. An Assoc Am Geogr 69(2):208–224
79. Sadeghi AR, Khakzand M, Jangjoo S (2019) Historical analysis of the role of bazaar on the formation of iranian islamic urban forms; case study: Shiraz, Iran. Armanshahr. Archit. Urban Dev 12(26):89–101
80. Souri E, Pakzad J, Foroughmand Arabi H (2020) Challenges of participatory urban design: suggestions for socially rooted problems in Sang-e Siah, Shiraz. Urban heritage along the Silk roads: a contemporary reading of urban transformation of historic cities in the Middle East and beyond, pp 247–259

81. Chahardowli M, Sajadzadeh H (2022) A strategic development model for regeneration of urban historical cores: a case study of the historical fabric of Hamedan City. Land Use Policy 114:105993
82. Mohammadpour S, Mehrjou M (2020) Evaluating the sustainability of urban neighborhoods with a low carbon approach; case study: Jowlan neighborhood of Hamadan. Commun Dev (Rural and Urban Communities) 12(1):21–50
83. Safayipoor N, Zarei J (2017) Community -based planning and urban renewal, sustainable worn tissue with an emphasis on social capital Case Study: Jowlan neighborhood of Hamadan. Geogr Plann Space 7(23):135–150
84. Etemad Sheykholeslami SF, Mofidi Shemirani SM, Mortezaei SR (2022) Recognizing the pattern of hamedan neighborhood center in accordance with the needs of women. Armanshahr Archit Urban Dev 15(38):169–184
85. Zaree ME, Pilehvar M, Eslami K, Alitajer S, Araghchian M (2020) Jowlan (Jowlan neighborhood, Hamedan). Talayi Publication, Tehran
86. Sajadzadeh H, Sharifi A, Soleimani S, Khanian, M (2017) Assessment of renovation and conversation planning of project mechanism in traditional neighborhood regarding to the environmental quality, case study: Julan neighborhood in hamadan, Iran. Armanshahr 17, 355–364. https://sid.ir/paper/202388/en
87. Alipour H, Thagha al-Islami O, Johari Timuri E (2013) Comparative study of Islamic Iranian identity in traditional and contemporary neighborhoods of Iran, case study: Fahadan neighborhood of Yazd and 3–2 neighborhood of New City of Binaloud, National Conference on Architecture and Human Urbanism Gera, Qazvin. https://civilica.com/doc/248806
88. Hosseini A, Fanni Z, Momeni A (2022) Comparative analysis of urban vitality in old and new neighborhoods with the approach of the right to the city a case study of neighborhoods in Yazd city. J Geogr Urban Plann Res 10(2):29–51

The Concept of Housing Transformation and Rehabilitation for Housing Development and Land Use Change in Biryogo, Kigali

David Mihigo

Abstract Healthy buildings are important as most people need well-constructed houses for their daily lives and durable houses to avoid accidents which may happen as well as the destruction of the houses. Therefore, this study examined the concept of housing transformation and rehabilitation in Biryogo Cell, City of Kigali for healthy buildings and climate-resilient cities. Interviews with 50 residents of Kigali city and One (1) Senior Engineer in the Infrastructure Department in Kigali city were done to capture views on the concept of housing transformation and rehabilitation as well as their impact on healthy buildings contrary to previous forms of old houses. Additionally, 30 site visits and observations have proven current development achievements. Results demonstrated that 91% Biryogo cell residents see housing transformation and rehabilitation as a helpful influence by giving an opportunity for durable houses, an increase in land value, and saving more land for additional uses. Housing development professionals responded that housing transformation and rehabilitation impacted suitable land uses and changes in the city. Despite these advantages, the concept is limited with insufficient financial capability to ease the idea of housing transformation and rehabilitation. Houses that result from housing rehabilitation and transformation are those located on asphalt roads in Biryogo cell. Car-free zones are also well-provided and connected to other roads including the main road from and to the City Center for smooth accessibility. Hence, this study will have much contribution on houses by increasing healthy buildings and climate-resilient cities in developing and developed world. These housing transformation and rehabilitation considered Four (4) components of sustainability which are human, social, economic, and environmental. The study recommends the concept of housing transformation and rehabilitation in informal settlements exposed to destruction for future climate-resilient cities.

Keywords Housing · Transformation · Rehabilitation · Housing development · Land use · Biryogo cell · Rwanda

D. Mihigo (✉)
Hangzhou International Innovation Institute of Beihang University (BUAA), Hangzhou, China
e-mail: mihigodavid7@gmail.com

A. Cheshmehzangi et al. (eds.), *Designing Healthy Buildings and Communities*, Urban Sustainability, https://doi.org/10.1007/978-981-96-3057-8_8

1 Introduction

House is a necessary and basic population's essential that each home necessitates entree. The fabric offers place, facilities, also comfort used for existing. Providing sufficient, cheap as well (house) is a fundamental purpose of each nation, in addition this escalated toward forefront in universal agendas over implementation of 2030 agenda for long term growth, by means of its committed cities Goals (SDG 11), and the novel urban agenda throughout 2016 United Nations Meeting on Housing and Sustainable Development (Habitat III). Housing touches good life of people and takes intense communal and financial influences on poplar's survives in each government [1].

As stated by UN, ninety percent of urbanization will happen in African countries in the next years. It is expected that 2 billion individuals will have life in informal settlements. Giving household earning less home and power accessibility, concerning quick urbanization, intensification of low-income neighborhoods, and climate change is urban areas burning issue [2].

Globally, extra than sixty-five million people are on change because of climate change effects or conflicts. Moreover, from villages into cities movement remains to escalate burden on built-up infrastructure and housing requests. Whereas this matter has persevered in the developing world and nations in evolution for years, current inflow of immigrants in Europe has set safe similarly affordable housing back on the Global North plan. At this point, the high pressure on cities and the disappointment of local government to offer access to inexpensive houses have directed to a situation where more than 1 billion city occupants are existing in under-serviced and informal neighborhoods globally [3].

After fifty years, the planet experienced an extensive increase in cities' inhabitants, starting around twenty nine out of a hundred of the globe's residents in 1950 to forty eight percent by 2003. Hence, the rapidity and progress of this increase origin main encounters, and checking these expansions and forming sustainable town growth environments keep on critical problems on the intercontinental advancement agenda [4].

Mass housing neighborhoods characterize biggest contribution of urbanity and morphological image of the big scale-cities in western-word, and the primary arrangement of urban transformation as well as extension in the second half of the twentieth century. Additionally, they have an important share of overall housing stock across Europe and partake a massive importance in guaranteeing admittance to reasonable and suitable housing, minimizing socio-spatial polarization in urban areas, and lastly contributing to social fairness [5]. Despite the hard work of local governments, civil society, Nations, the international community, and private sectors, accessibility to adequate housing remains a key challenge for a growing number of people all over the world. But what is possibly different recently is that this challenge is distressing people within countries at all levels of economic and social development [6].

In developing countries, for instance in Brazil, housing transformation takes various systems and challenges occur to promote low-income families' living standards. The minimization of housing scarcity is the major aim of government housing Programs, which frequently have political and economic objectives and consequently favor new constructions [7].

Bangladesh, due to its exceptional geographical location, is extremely exposed to consistent and numerous ordinary hazards, counting landslides, overflows, drought, storm flows, tropical cyclones. Therefore, these threats, united with an enormously high people density as well as poor socio-economic circumstances, are now top to the limited and complete devastation of housing, land, and property, the damage of livings as well as extensive migration and movement across the country. Additionally, Bangladesh is amongst countries utmost exposed to the impact of climate change [8].

Generally, this study is aiming at highlighting the idea of housing transformation and rehabilitation for housing development and land use change in Biryogo cell, city of Kigali, the country of Rwanda, where the study results will inform future city developers not only in Kigali but also in both developing and developed countries in transforming and rehabilitation their old houses in their respective countries besides healthy cities. Very closely linked to four components of sustainability consideration that are (human, social, economic, and environmental). Thus, the following were specific objectives of this study on the concept of housing transformation and rehabilitation for housing development and land use change in Biryogo: (1) To demonstrate the recent situation of housing transformation and rehabilitation for housing development and land use change of Biryogo and healthy buildings and increase of building-level practices in developing countries; (2) To highlight some good concepts adopted in Biryogo during settlement or houses transformation and rehabilitation idea for healthy buildings and climate resilient cities; (3) To recommend appropriate concept for housing transformation and rehabilitation for housing development, healthy buildings and climate resilient cities.

Furthermore, this chapter is organized into five sections and section one is 'Introduction'. This part gives the background information of the study and literature review related to the study. Section two is 'Materials and Methods'. This section also offers an overview of the study area as well as data collection procedure and analysis. Section three is the 'Results'. This section details the results from the data collected on the field including specific group of individuals of the interviewees, the concept of housing transformation and rehabilitation for house development in Biryogo cell, Kigali city such as the future of housing transformation in emerging technology. Section four is the 'Discussion', where the concepts from other researchers mentioned similar information the same as information used in this study are demonstrated. In simple words, this section is used to mention what other researchers said on similar topics under study. Finally, section five is the 'Conclusion', where the study highlights the necessary information that concludes the report.

2 Materials and Methods

2.1 *Overview of the Study Area*

Government of Rwanda is made of two layers of administration (local, central) and 6 managerial bodies: These constructions, were restructured below the 2005 change, are corresponding. Rwanda is separated into 4 provinces as well as Kigali city which are alienated into 30 districts. Those districts are divided into 416 sectors. Those sectors are divided into 2148 cells (Akagali) which are also divided into 14,837 villages. In addition to this, all these parts are represented by various persons at each level and they play diverse roles [9]. Kigali is the capital city of Rwanda, geographically located at the heart of the nation with an area of 730 Km^2, the city is located at longitude 30°07' E and latitude 10° 58' S.

The normal precipitation rate is almost, with a yearly regular great temperature of 27 °C and an annual average low temperature of 16 °C. The city of Kigali is made up of three districts namely Nyarugenge, Kicukiro, and Gasabo. Thus, Gasabo is the largest district with an area of 429.3km^2, followed by Kicukiro (166.7 km^2) and Nyarugenge (134km^2) [10]. According to [11] the 5th (RPHC5) city of Kigali's population is 1,745,555. The country of Rwanda has experienced resilient and constant economic growth over the last two decades after the 1994 Genocide. In Africa, Rwanda is among the fastest-growing economies. In 2010 Rwanda's per capita income improved from 587$ to 778$ in 2018. Economic development was around 7.2% per year during the same period compared to the regional reached 9.4% in 2019. In Sub-Sahara Africa, this was the second-highest rate of economic growth [12] (Map 1).

2.2 *Data Collection Procedures and Analysis*

This part clarifies different research tools that were adopted during information assembly. Therefore, secondary and primary data were considered in this research where secondary data were putted together by reading literature associated with healthy building and climate-resilient cities. The primary data were obtained by observations and consultations. Field observations were used to see recent situation, building-level practices, housing transformation, and rehabilitation in Biryogo. Photos were taken to stipulate the housing transformation happening in the study area. 50 Nyarugenge District residents together with one senior engineer in the Infrastructure Department in the city of Kigali were interviewed concerning the idea of housing transformation and rehabilitation for housing development and land use change in Nyarugenge District specifically in Biryogo to capture their views on the concept of housing transformation and rehabilitation for healthy buildings and climate resilient cities.

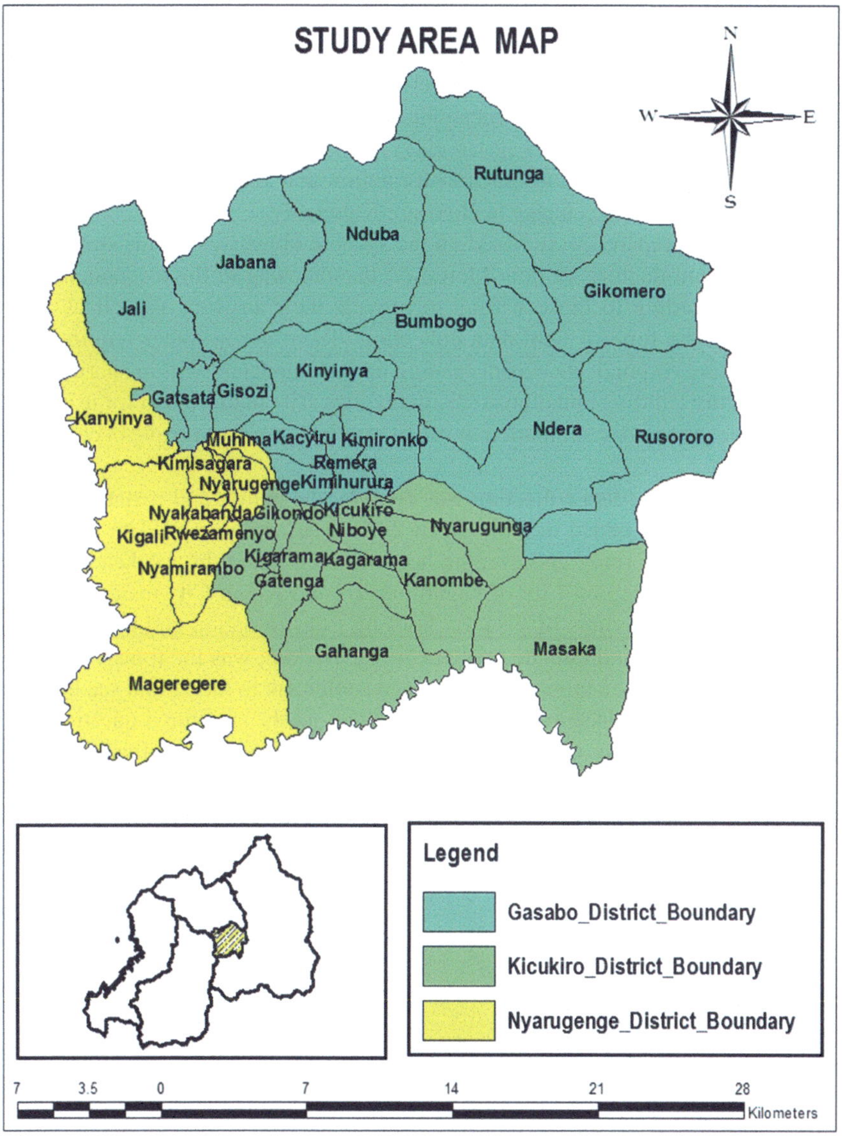

Map 1 Study area map. *Source* Author (2024)

The key aim of selecting a single expert was to have his point of view on healthy buildings and climate-resilient cities. Thus, one expert from the Infrastructure Department in Kigali city was sufficient because he represents other field and professional experts in the Infrastructure Department. After all, they share a mutual understanding of what can be done for the city and future activities the city must have to keep healthy buildings and climate-resilient cities. This is explained by community participatory planning which offers the room for experts to discuss together the better approaches to elevate building-level practices such as innovative approaches and housing transformation to withstand the climate-related effects. Having expert opinion in this study enriches the results. Furthermore, questionnaires were structured for data collection purposes and this was proper way for every person's perspectives about housing transformation concept of housing transformation and rehabilitation for housing development and land use change in Biryogo, Kigali city.

All questions prepared were based on the specific objectives of this study highlighted in the introductory section. Moreover, the choosing of the respondents was established according to those willing to provide their answers as well as those present during the time of field data collection. There are no other requirements used in selecting respondents except random and respondents' will in sharing their ideas on healthy building and climate resilient cities. The main reason for using this tactic is to escape any discrimination or use of a certain group of people in collecting answers.

In addition to this, data collection was done by the author and it started on 7th September 2023 and ended on 30th April 2024. The activities of data collection were done during morning and evening hours as well as on the weekends mostly on Saturdays depending on City of Kigali experts and Kigali City residents' presence in sharing their points of view on housing transformation and rehabilitation for healthy buildings and climate resilient cities. Walking was the transport means the researcher used when crossing the Biryogo settlement to locate and see housing transformation taking place in Biryogo then filling out the structured questionnaire before data collection protocols. Gathered data were analyzed using Microsoft Excel for quantitative data analysis and Microsoft Word for production and summary of this report. Finally, regarding analysis, the study adopted descriptive statistics such as figures and tables for quantitative data while qualitative data were analyzed by using content analysis.

3 Results

3.1 Socio-Demographic Characteristics

This section on respondent's data was studied to mention respondents who shared their valuable time and contributed to the interview by offering the answers to the structured questions of this study. Hence, Table 1 summarises information concerning participants in the interview. From education viewpoints, those who joined University education were 70% then 12% attended secondary education while the remaining 18% studied primary school in enlightening the concept of housing transformation and rehabilitation for housing development and land use change in Biryogo cell and for healthy building and climate resilient cities. Regarding Gender in this study, women participated in the interview more than men with 69% while men were

31%. The inclusion of both in this study was good as both men and women must have equal opportunity in providing their point of view in shaping the future of housing transformation and rehabilitation in cities and at the same time improving the building level practices. Based on participants' employment activities, a large number were from the private sector with 71% of self-employment, mainly doing business, tailoring, mechanical, and many more jobs. Then 13% were from the public sector serving various government institutions.

This is because, during data collection, private sector participants were present more than public sector workers due to being most of the time to work, especially when data collectors received responses. The remaining 16% were unemployed people, this shows that unemployed persons took second place after the private sector simply because unemployed people were likely to be available comparatively to public sector personnel. Age was another aspect identified in this study where people the age between 30–40 came first (more than 62%). While the respondents aged between 40–50 followed (with 25%). Then 8% of the respondents in the age range between 18 and 30. Then the age range above 50 years old was represented by 5%. Having age differences offers room for better answers than having a single or two age ranges in receiving responses from respondents.

From the results people with the age range 30–40 years have much impact in housing transformation and rehabilitation because they are the ones to own them or sell them. In addition to this, the author decided to add the opinions of people from the age of 18 because he wanted to collect youth perspectives in terms of housing

Table 1 Socio-Demographic characteristics

N°	Respondent	Classifications	Frequency	%	Total
1	Sex	Male	16	31	50
		Female	34	69	
2	Age	Age 18–30	4	8	50
		Age 30–40	31	62	
		Age 40–50	13	25	
		Age above 50	2	5	
3	Education	Primary Education	9	18	50
		Secondary Education	6	12	
		University Education	35	70	
4	Employment	Private Sector	36	71	50
		Public Sector	6	13	
		Unemployed	8	16	
5	Marital Status	Single	30	60	50
		Married	12	23	
		Divorced	8	17	

Source From Fieldworks conducted by the Author (2024)

transformation and rehabilitation for healthy buildings and climate-resilient cities. It also concurs with aged people' opinions to contribute to the study for better results of the study.

3.2 *The Concept of Housing Transformation and Rehabilitation for House Development in Biryogo*

3.2.1 The Concept of Housing Transformation in Biryogo

The action of housing transformation and rehabilitation is happening nowadays in various areas in the city of Kigali, but Biryogo is exceptional because most of the houses were old and needed rehabilitation or transformation instead of being removed. Figure 1 demonstrates a commercial house in Biryogo, Nyamirambo transformed into a large commercial Block with more than six rooms which currently serves different business purposes as you can see in the figure below. Before the houses were old enough with few rooms but recently transformed houses increased their value as well as durability for healthy buildings and climate-resilient cities and improved the beauty of the city. This was confirmed by 87% of the respondents that housing transformation increases the value of the house at the same time rooms for rent. Additionally, the house is constructed near the major road (asphalt road) from the city center to Kuryanyuma or other places such as Gikondo, Remera, and Nyabugogo. A drainage system for rainwater and unwanted water is around the houses to drain them to designated outlets.

Building-level practices in Biryogo are attaining another level compared to the time before this one where vertical housing transformation was taking place when rich people came into this settlement negotiated with local people their property then paid them so that they elevate a good building storey which will accommodate many activities including commercial as you can see in Fig. 2. A new building storey was constructed recently where old houses constructed as an example to the next side of that building storey. It is not only this building but many building stories that are being constructed in Biryogo where most of them will serve commercial purposes. This testifies to healthy buildings and climate-resilient buildings in Biryogo because nowadays constructed building storey and transformed horizontal houses stand well against climate-related problems, boost their durability, increase the value of land as well as offer the city beauty. 98% of Respondents attest that transformed buildings offer enough rooms for rent, such as offices and business rooms for them to conduct their development activities. The good thing about this settlement is connected to an asphalt road for people to move around, in and out of the settlement. From Fig. 2 property owner is able to generate more money from rent, as this building storey has many rooms exposed for rent. Apart from climate resilient building the building offers a secure and comfortable place for working.

Fig. 1 Housing transformation into large commercial house. *Source* The Author (2024)

Fig. 2 Emerging vertical Housing development in Biryogo. *Source* The Author (2024)

3.2.2 The Concept of Housing Rehabilitation in Biryogo

To reach high level of healthy buildings and climate resilient city housing rehabilitation is a key to open this future. Because housing rehabilitation aims at improving house health as well as the beauty and comfortable place for people to live. In Biryogo, house rehabilitation is also taking place for many houses and nowadays you can see that different houses are being rehabilitated, when for example house owners decide to paint their houses with modern paints and weather guards' paints to resist on the rain and other climate associated negative effects. Figure 3 shows rehabilitation works in Biryogo, when housing became beautiful, climate resilient and increased their value. Housing accessibility to major roads was improved to make sure vehicles and pedestrians can go effectively and without rain water on the top surface of the soil or pavement. Road signs were provided on the road to direct movement of vehicles, cycling and motorcycles for not to face accidents on the road or neighboring communities. 85% of the respondents said that housing rehabilitation in Biryogo advanced housing development for property owners as well as for healthy buildings and climate resilient city in Kigali.

Most of the time, country and urban development are based on basic or public infrastructure. That is why the recent situation of public infrastructure in Biryogo is important to highlight for a better future for cities as well as healthy settlements and

Fig. 3 Housing rehabilitation in Biryogo. *Source* The Author (2024)

Fig. 4 Public infrastructure provision in Biryogo. *Source* The Author (2024)

climate-resilient cities. Figure 4 proves the current situation of Biryogo's infrastructure development. Where the key major roads that connect various areas of Kigali city were constructed enough and in asphalt form, for smooth automobile movement. Walkways or pedestrian ways were provided too, for pedestrian's safe walking along the road. Electrical connection cables are placed within the settlement for homes to connect to the electricity and then become secure in their homes. In addition to these street lamps, they are installed on the road to increase security even facilitate pedestrians to walk safely and anytime they need that even during night later hours. This also goes hand in hand with clean water distribution within the settlement for community well-being. For instance, 92% city of Kigali residents highlighted that basic or public infrastructure in Biryogo is sufficient for people to live in this place while enjoying these infrastructures development as well as this transformation and rehabilitation happening in Biryogo.

3.2.3 The Concept of Car-Free Zones in Biryogo

It is a very impressive concept to have car car-free zone known as (Ku Marange) where people, both local and international communities, come to look for Tea, coffee, or food at affordable prices. This idea helps a lot, especially during pandemic times like COVID-19 which happened globally where most people were advised to use outdoors rather than the indoor of houses, according to WHO regulations because inside the building the pandemic spreads more than outside the house. Using outside the house helps the community enjoy the city view and fresh air. This concept gave the attractiveness of the city of Kigali as well as Biryogo where this place nowadays is the most visited area in Nyamirambo Sector with people from all corners of the city.

The cost of drinking and food is not expensive, it is moderate and affordable for poor communities to buy them. It is not only that but also old houses reduced above 75% because people changed their roofs and walls from soil bricks into cement blocks and clay bricks. In that free zone area roads are painted and cleaned accordingly. Tents are movable just to protect people from strong sun and heavy rain which might come and affect them. House owners especially those constructed near the road were informed by the city of Kigali to transform their commercial houses to meet commercial housing standards and to become strong enough for any accidents that may arise due to unstable houses or doing business in unsuitable houses which can facilitate the occurrence of both communicable and non-communicable diseases.

This also boosts healthy buildings in the city and has climate-resilient strategies in practice. Trees are planted at that place along the road and these trees are growing day by day to provide fresh air for people sitting under them. Car-free zone roads are made in asphalts and then painted with paints in white color, green, and blue colors as shown in Fig. 5. For cashless transactions, mobile services are in that zone to ease clients' operations and to avoid hand payments and money exchange rather than digital payment and buying during any diseases that might come soon. 100 respondents confirmed that this type of car-free zone was needed to fulfill the needs of the community not only in Biryogo but also across the country, even those coming from the countryside or villages in Rwanda as well as foreigners from different countries at the global level. Car car-free zone areas are not only for the attractiveness of the city or place for eating and drinking because even many houses were rehabilitated and transformed in line with the idea of open space. These could also help to resist the climate-related actions that may occur at any time.

This place was designed for children's use, but sometimes even old people use this place, especially during walking and playing games with children. Again, this area is famously known as "Amarange y'abana" where Children can go and play different games such as football, running, and many more. This place welcomes many children including those with disability while interacting with their fellow children who do not have any disability. Figure 6 illustrates car free zone in Biryogo left for children's relaxation and play in case they need it. Cars and motorcycles are prohibited from reaching this place to avoid any accident which may happen and cause the death of children. 97% of respondents said that this place was needed and useful for their children as well as old persons who need to sit there and relax or have conversations with their friends or their neighbors. As you can see in Fig. 6, the place is having electric lamps which easy walkability during the night or other activities that may take place during the night in the area. Benches are also present to facilitate those who need to sit. Pavers are laid on the surface to elude stagnation of the rain water on the surface which can stay on the top surface and stop children's games.

Fig. 5 Car free zones in Biryogo neighborhood. *Source* The Author (2024)

3.2.4 The Future of Housing Transformation in the Digital Era

Digital Technology is a topic under debate by different organizations and companies that deal with housing development, and urban planning in cities. Based on this mentioning the future of Biryogo in the digital era was necessary as city planning is

Fig. 6 Car-free zones for children in Biryogo settlement. *Source* The Author (2024)

transforming into this trend for increase their efficient work, optimal public services utilization as well as early alerts on city's problems such as fire attacks. Therefore, in Biryogo some buildings are being transformed and are considered digitization tools to make their building safe and ready for externalities that might arise accidentally. It is for this reason Cameras to monitor the building day by day were installed on buildings there in Biryogo so that they give necessary information on building situation even when the building owner is not around.

Some Cameras are connected to Smart Phones for property owners to check their houses daily on their mobile phones without going on the site to check everything. This gives hope that as rich persons transform the settlement this will reach the level where many buildings will be connected and be able to share key information for smart housing management this also will help the government because it will be easier for the government to implement smart housing and smart city concept while having necessary information in hand from various houses across the city. This was agreed by 78% of the respondents that technological tools such as Cameras are useful to monitor the building, especially in the absence of the building owner. Instead of being there to check everything and going on his or her building. Figure 7 determines the constructed building in Biryogo which used a Camera to monitor the building. This also helps in achieving healthy building and climate-resilient cities because technology will reveal climate-exposed areas and experts respond in time before the problem happens.

Fig. 7 Gentrification and digitalisation in Biryogo. *Source* The Author (2024)

3.2.5 City of Kigali Professional Perspective on Housing Transformation and Rehabilitation

Kigali city expert from the infrastructure department and civil engineering division, the construction regulations must be simple for the easy construction process in the city as well as the development of a masterplan to regulate construction projects in cities because some people tend to construct their houses that are outside master plan concept and this is not good because it creates informal development and delays service infrastructure distribution in cities. He added that ring road construction are better approach in towns and cities to facilitate accessibility in transformed and rehabilitated housing in urban areas. Among the challenges he mentioned is budget which can solve or speed up the concept of housing transformation and rehabilitation for healthy buildings and climate-resilient cities. Transformation of houses also is associated with land use change where most of the time residential houses change to commercial houses and this is applied mostly to houses located near the major roads as well as the secondary roads.

4 Discussion

Transformation and rehabilitation of houses in the city of Kigali are being done in the matter of adding the durability of houses without forgetting the city's good and drainage of rainwater in a proper way which will not affect neighboring houses and cause erosion or flooding within the settlement.

Housing transformation, rehabilitation plays critical role in city's development and management once they become good enough to accommodate inhabitants in comfortable places that can resist climate change actions including flooding and land sliding which is occurring currently in various cities. So-called gentrification in urban areas where someone with financial capacity comes into an informal settlement or neighborhood with low-income housing development and then buys one to three houses and then removes those houses for him or her to get enough space for his or her building storey.

Furthermore, this building might have up to six stories which gives more functionalities like business rooms, rent rooms, office rooms, and parking areas for business owners and visitors at the building. Jobs are created during the construction of that vertical development where steelworkers and masons get employed and work for their personal development from the money, they receive from that construction project. Once this building is constructed, it can provide more services to neighboring communities and in a small area, also drainage of both unwanted and rainwater can be done properly in a way that can't affect people's properties. Rehabilitation again can make houses stable and durable compared to the time that houses could stay due to the use of new and durable construction materials.

When arriving in Kigali city, Biryogo cell you will be able to observe these changes taking place there especially residential houses being transformed into commercial houses where citizens are trying their best to make this area better than before instead of relocating themselves to other places. Residential houses are also being transformed from horizontal development into vertical development or from old horizontal housing development into horizontal transformed housing development. This varies depending on housing owners' financial ability.

Additionally, housing transformation and rehabilitation increase the rent cost of the house due to various construction materials added to the existing old house including paints, concrete, reinforced concrete, bricks, and concrete blocks. Apart from that, even the cost of the house increases. Initially, a house may have two commercial rooms but after some changes, this house can be transformed into 4 business rooms. Thus, most of the room rent per month is 100,000 thousand or 150,000 thousand Rwandan Francs depending on the location of the house as well as how big is that room. According to [13] cities are not areas where only people work, they are places where they consume goods and services as well as play.

However, the development of cities creates costs such as the cost of housing, and pressure on infrastructure. Thus, this puts it at the top of the best city organization and has the capability to successfully plan for growth. In developing countries housing transformation can be a suitable solution especially for rapidly growing cities including mega cities for instance Dar es Salaam and Kigali because most of the people do not have enough financial capacity to remove existing development for someone to construct new buildings. Some areas can be reserved for planned developments but informal settlements on the other hand for cost saving community can do transformation according to planned transformation as well as agreed changes which can be done on a certain house depending on the land uses and location of that house. Housing modification increases the durability of houses if the houses were to last for five years, after transformation they can last for more than five years, this also is healthy for building and climate-resilient neighborhoods. It is good for families in African countries not to construct another house by starting from excavation, foundation up to the roofs.

5 Conclusions

This study checked the concept of housing transformation and rehabilitation for housing development and land use change as well as for healthy buildings and climate-resilient cities. Housing transformation and rehabilitation are mostly adopted for old houses or information settlements to withstand the climate-related effects and stay long for houses' beauty and durability. Though the achievement of healthy building in these neighborhoods is affected by a lack of sufficient financial capacity to transform many houses in the cities and escalate building-level practices in urban areas, by defining the objectives of this study, some indicators to evaluate housing transformation and rehabilitation for healthy buildings and climate resilient cities are the following: (1) Existence of transformed and rehabilitated houses in the neighborhoods and cities; (2) Healthy buildings and climate resilient cities took into consideration digital technologies like Internet of Things (IoT); (3) Transformation and rehabilitation activities taking place daily in cities or settlements.

Finally, future urban researchers can advance the results of this study by adding many studies in addition to this study mostly topics that were not mentioned in this study to supplement this study and highlight the key information for healthy buildings and climate-resilient cities. This goes hand in hand with international suggestions when future housing developers, urban planners, decision-makers, and international researchers consider the concepts revealed in this study that were not present in recent housing rehabilitation and transformation in cities. Such as [8] *Use of emerging technologies for managing transformed buildings and the entire settlement*; [4] *Settlements or Neighborhoods building and housing transformation where they didn't take into consideration of Open space as demonstrated in Biryogo place known as Ku Marange as well as Amarange y'abana must initiate to work on this concept for healthy cities*; [10] *Use of Construction materials that are available in the settlements or within the city that are cheaper and strong to increase durability of transformed and rehabilitated houses.*

Hence, after conducting this study of the concept of housing transformation and rehabilitation for housing development and land use change in Biryogo, the city of Kigali for healthy buildings and climate-resilient cities the study suggests some recommendations for a better future of neighbourhood rehabilitation and transformations in cities as well as building-level practices improvements. Informal house transformation and rehabilitation is a recommendation that can increase healthy buildings and boost the climate-resilience of houses and stand long without other externalities. On one hand to the other neighborhood transformation or rehabilitation can be a good tool to change houses' appearance and durability instead of expropriation which is expensive for countries while shifting communities from their living places to other locations.

Additionally, this affects community social cohesion once moved. Social Cohesion maintaining while avoiding community relocation is very important as it improves community lives and development based on their social relations and interaction within the settlement. During construction materials transportation and house rehabilitation, the use of local, cheaper, and durable construction materials is another recommendation for an economical way of housing transformation and rehabilitation.

Acknowledgements The urban researcher would like to thank the City of Kigali expert from the infrastructure department and civil engineering division who took his time and shared his perspective on healthy buildings and climate-resilient cities in developing countries. Moreover, Kigali city residents for their valuable time spent with the researcher in contributing to this study while giving their responses during interviews based on structured questionnaires for collecting necessary information on building-level practices enhancement as well as achieving climate resilient cities in African states.

References

1. Moreno-Monroy A, Gars J, Matsumoto T, Crook J, Ahrend R, Schumann A (2020) Housing policies for sustainable and inclusive cities: how national governments can deliver affordable housing and compact urban development. Coalition for urban transitions. OECD and Coalition for Urban Transitions, London and Washington, DC Retrieved from https://Urbantransitions.Global/Publications
2. Sunikka-Blank M, Abdie D, Bardhan R (2021) 'We need ground space': urban densification and transitional housing in Ethiopia. JBA 81–106. https://doi.org/10.5871/Jba/009s9.081
3. Ley A, Rahman MA, Fokdal J (2020) Housing and human settlements In a world of change. Habitat-International, International Urbanism (Städtebau-Institut). The Deutsche Nationalbibliothek, Berlin. Retrieved April 23, 2024, from https://doi.org/10.14361/9783839449424
4. Jenkins P, Smith H, Wang YP (2007) Planning and housing in the rapidly urbanising world. Taylor & Francis Group. Routledge, USA and Canada
5. Milovanović A, Dragutinovic A, Nikezić A, Pottgiesser U, Stojanovski M, Deskova AI et al (2022) Rehabilitation of mass housing as a contribution to social equality: insights from The East-West European academic dialogue. Sustainability 14:33. https://doi.org/10.3390/Su14138106
6. Un Habitat (2021) The role of land in achieving adequate and affordable housing. United Nations Human Settlements Programme (Un-Habitat), Kenya. Retrieved April 23, 2024, from www.Unhabitat.Org
7. Kowaltowski DC, Muianga EA, Ruschel RC, Silva VG, Granja AD, Moreira DD et al (2023) Housing Transformation in search of well-being. Xvi Encontro Nacional De Conforto No Ambiente Construído, 11. Retrieved from https://Www.Researchgate.Net/Publication/369907843
8. Arteria Studio (2014) Bangladesh Housing, Land and Property (Hlp) rights iniative. Displacement Solutions & Young Power In Social Action (Ypsa). Arteria Studio, Bangladesh. Retrieved from Arteriastudio.Com.Au
9. Republic of Rwanda (2024) Administrative structure. Retrieved April 20, 2024, from Republic of Rwanda. https://Www.Gov.Rw/Government/Administrative-Structure
10. Kigali City (2024) About Kigali City, April 20. Retrieved from Kigali City: https://www.Kigalicity.Gov.Rw/About/Overview
11. Nisr (2024) Key figures: 5th Rwanda population and housing census (Phc), April 20. Retrieved from National Institute of Statistics of Rwanda. https://www.Statistics.Gov.Rw/Publication/Key-Figures-5th-Rwanda-Population-And-Housing-Census-Phc
12. The World Bank (2020) Rwanda urban development project Ii (P165017). The World Bank, Kigali. Retrieved April 19, 2024
13. Sadlier R (2015) Using land for housing. The New Zealand Productivity Commission, New Zealand. Retrieved from www.Productivity.Govt.Nz

Healthy Buildings and Climate-Resilient Cities: The Case of Nantong City as Community-Level Interventions

Francis Deng Clement, Shutian Zhou, and Armin Mehdipour

Abstract In terms of constructing healthy buildings and improving climate resilience, Nantong City, located in Jiangsu Province, China, presents several challenges and opportunities. Signs of surface faults associated with land subsidence have been noted in the city, suggesting the presence of geological challenges. Urban residential land use fragmentation has also been a problem in Nantong, leading to land waste and environmental degradation. In contrast, other cities in China, such as Changzhou and Yantai demonstrate different levels of resilience, indicating that Nantong has been assessed to have a modest level of resilience. To handle these challenges and improve climate resilience, it is crucial for Nantong to concentrate on aspects such as economic support capacity, social development capacity, endowment of natural resources, technology adaptation capacity, and risk management capacity. The importance of forward-thinking adaptation planning to improve the total adaptive capacity of the city system is vital, as is the government's responsibility to lead the construction of climate-resilient cities. Essential measures include bolstering funding for urban climate change programs, enhancing city form and structure, and boosting Nantong's resilience to climate change. A better understanding of the spatiotemporal features of urban development and an assessment of the effects of climate change on population urbanization can shed light on how to create resilience and achieve sustainable growth. Nantong could improve its climate resilience and address the challenges of climate change by using resilience assessment and planning models and methods. To construct healthy buildings and improve city resilience in the face of ongoing environmental and climate change, Nantong City could benefit from a holistic approach that incorporates geological considerations, urban development strategies, and rate evaluations.

F. D. Clement (✉) · S. Zhou
School of Arts (Architecture), Nantong University, Nantong, China
e-mail: francisdeng@ntu.edu.cn

A. Mehdipour
Faculty of Science, Engineering and Technology, The University of Adelaide, Adelaide, SA, Australia

A. Cheshmehzangi et al. (eds.), *Designing Healthy Buildings and Communities*, Urban Sustainability, https://doi.org/10.1007/978-981-96-3057-8_9

Keywords Healthy buildings · Climate-resilient cities · Nantong City · Land subsidence · Urban residential land use fragmentation · Resilience assessment · Adaptation planning

1 What Is "Healthy Building"?

The term "healthy building" refers to a relatively new field aimed at enhancing the mental, emotional, and social well-being of occupants. The National Human Activity Pattern Survey reveals that Americans spend 87% of their time indoors and 6% in automobiles. Building on green building practices, the concept of healthy building emphasizes occupant health and efficiency, as well as environmental and economic impacts. Key benefits of healthy buildings extend beyond the immediate physical environment. These benefits include a noticeable reduction in absenteeism, with healthier indoor environments contributing to fewer health-related absences. Additionally, the design of such spaces encourages increased physical activity, which supports overall well-being. This leads to improved performance, benefiting individuals and organizations collectively. By cultivating a healthier and more active workforce, organizations can significantly enhance productivity, elevate engagement levels, and improve overall job satisfaction. The concept of healthy building integrates knowledge from construction, health, and building science, requiring collaboration among facility managers, architects, engineers, health specialists, and public health experts. Charrettes are an effective method for these multidisciplinary teams to collaborate, generate ideas, and develop solutions.

Indoor air quality is crucial for good building design, since poor air quality can worsen asthma, asbestosis, and lung cancer. Building materials, furnishings, and supplies can release chemical pollutants through outgassing, which negatively impacts the health, safety, and productivity of occupants. Increasing ventilation rates can enhance indoor air quality by dispersing pollutants and improving performance, but natural ventilation should be managed separately from thermal conditioning to ensure comfort. Pollutants, pesticides, volatile organic compounds (VOCs), lead, and dirt present potential health risks. High-efficiency filter vacuums can effectively remove allergens and dander that may irritate the respiratory system. In American culture, the widespread use of disposable products, especially in healthcare, can have adverse environmental effects. For example, disposable cotton towels produce 6.6 kg of CO_2 equivalents and 0.024 kg of nitrogen emissions, potentially harming the environment. The thermostat, clothes, atmospheric temperature, relative humidity, mean radiant temperature, and metabolic rate affect thermal comfort. An appropriate thermal environment meets the comfort demands of at least 80% of occupants, according to ASHRAE. Indoor relative humidity should be maintained between 30 and 50% to prevent condensation, with adequate drainage and ventilation to manage moisture. Excess moisture can lead to mold growth and deteriorate indoor air quality. In addition to atmospheric pollution, noise exposure can negatively impact stress levels, blood pressure, and mental health. The World Health Organization recommends developing a National Plan for a Sustainable Noise Indoor Environment to address these concerns. Moreover, water quality degradation poses a significant health risk, particularly in developing nations and water-vulnerable populations like children, highlighting the urgent need for stringent water quality regulations.

1.1 Why Are Climate-Resilient Cities Significant? How Do We Build Resilient Cities?

According to the Organization for Economic Cooperation (OECD), the number of people living in cities and urban regions is predicted to rise from 55% in 2024 to approximately 70% in 2050, with 1.4 million people arriving in the world every week. Cities face numerous challenges, such as the rise of electric cars, environmental degradation, economic shocks, migration, and crime due to increased population. Today's cities and communities are already feeling the consequences of both natural and human-made challenges: on the one side, there are political and governance failures, economic challenges that lower people's quality of life, and human conflicts and wars; on the other side, natural disasters like droughts, floods, and earthquakes have affected over 220 million people and caused an annual economic loss of $100 billion. These challenges have bleak future projections. Although it seems an enormous undertaking, the International Panel on Climate Change (IPCC) has urged governments and industry to keep global warming at 1.5 °C and 2 °C.

Extreme climate change events will happen more frequently and severely if we are unsuccessful in our mission to curb global warming. The World Bank predicts that 77 million people may return to poverty by 2030, and natural disasters could cost the world over $300 billion, if cities do not become more resilient by 2030. But if cities do succeed in becoming more resilient, how can they deal with these urban challenges? The central theme of World Cities Day is the construction of cities that are both environmentally sustainable and capable of withstanding challenges. Resilient cities can effectively manage disasters brought on by both natural and economic causes, to protect life and property, and to reduce environmental and social risks. The United Nations recommends diversifying the economy, creating business and employment opportunities, and engaging with the private sector for economic protection. Cities should be planned to minimize environmental impact, strengthen infrastructure, and focus on public education for climate and environmental resilience. Educating citizens about democratic, sustainable, and inclusive decision-making can foster social cohesion.

Why focus on cities? Urban areas play an important lead in global sustainability due to their rapid growth and heightened energy demand, which results in a substantial carbon footprint. By 2030, it is projected that over 60% of the global population will reside in urban areas. The Industrial Revolution and economic globalization have hastened the growth of urban opportunities, yet not all cities have made full use of these. Cities with a strong desire to experiment and innovate are more likely to adapt quickly. The capacity to experiment and innovate is the foundation of hope and fortitude, and cities need to conduct new experiments in green urbanism to overcome the fear of change. Urban areas in the United States are frequently hubs of exceptional leadership and creativity. "*The Mayors Climate Protection Agreement and the C40 Cities Climate Leadership Group highlight the emerging leadership from government, business, academic institutions, and community organizations. There are inspiring stories of hope in cities worldwide, especially in American cities that are in urgent need of development*".

It is worth discussing how resilient cities can contribute to addressing the challenges of climate change and declining oil availability. Resilient cities offer various benefits, including improved physical and mental health among the community, efficient resource use, and better accessibility. However, many cities lack comprehensive models to reduce oil dependence, focusing instead on population growth and general sustainability. Certain cities, such as New York and San Francisco, have commenced efforts to mitigate greenhouse gas emissions and recognize peak oil; however, comprehensive strategies are frequently lacking. It is also worth highlighting the reliance that rural areas have on oil, especially in industries like agriculture and tourism; this raises concerns about the sustainability of rural productivity. Environmental advocates have pointed out current resource consumption issues but often struggle to propose clear, positive alternatives. This chapter aims to fill that gap by providing solutions for cities and rural areas to adapt to these environmental challenges, taking Nantong City as a case study.

2 Background: Healthy Buildings and Climate-Resilient Cities Focus on Nantong City

An analysis of literature regarding healthy buildings and climate-resilient cities, particularly in Nantong City, provides valuable insights into urban resilience against climate challenges. The notion of climate resilience is pivotal, highlighting urban capacities to withstand and adapt to climatic disturbances. Healthy buildings play a pivotal role in this resilience, as they go beyond structural soundness to address the well-being of their occupants. Features such as improved air quality, natural lighting, and effective thermal regulation contribute significantly to reducing the adverse health effects of extreme weather. In Nantong, incorporating these elements into urban design ensures that buildings remain functional and supportive during climate events like heat waves and floods.

The incorporation of green infrastructure, including urban parks and sustainable water management, has become a significant on enhancing urban resilience and improving the quality of life for residents. By integrating these elements into city planning, communities can better manage stormwater, reduce urban heat, and create inviting spaces that foster social interaction and well-being." These approaches alleviate urban heat island effects and improve urban resilience to climate change. In Nantong, where rapid urbanization is constant, green infrastructure offers a dual benefit: it creates more liveable spaces while bolstering the city's capacity to respond to environmental stressors. Energy efficiency is another crucial aspect. Retrofitting existing buildings with sustainable technologies reduces energy consumption and makes urban environments more resilient to disruptions. These upgrades enhance resilience by reducing reliance on external energy sources. Finally, effective governance and community involvement are indispensable to building climate resilience. Engaging residents in the planning and decision-making processes allows cities like Nantong to develop strategies that are both locally relevant and broadly effective. Localized governance structures that facilitate this participation contribute to stronger, more responsive urban systems that can better address the multifaceted challenges posed by climate change. Through a combination of thoughtful urban planning, technological innovation, and active community participation, cities like Nantong can not only survive climate shocks but also emerge stronger and more sustainable in their aftermath [1].

Researchers have stressed the significance of evaluating, planning, and constructing climate-resilient urban areas, emphasizing essential features, strategic planning, and assessment methodologies [2]. Urban resilience frameworks, such as the urban resilience scientific framework, are designed to improve the responses of cities to climate change by incorporating the complexities and specificities of the specific location [3]. Assessing the resilience of urban areas to climate change is essential as it lays the groundwork for the creation of tools, such as urban resilience indices, that are used to monitor progress towards climate-resilient cities [4]. Considering the substantial influence of climate change on urban environments, it is imperative to improve the capacity of cities to withstand and adapt to climate change [5]. Monitoring the effectiveness of urban climate resilience and adaptation initiatives is critical and the use of indicators is essential for this purpose [6]. However, a multidisciplinary approach is needed cities [7]. Furthermore, urban gardening is gaining global recognition for its crucial role in enhancing a city's ability to endure the impacts of climate change [8]. Investing in climate-resilient infrastructure in urban areas is essential to preventing the costs associated with climate disasters and facilitating swift recovery from such events [9]. Legislation can play a crucial role in fostering urban climate resilience by establishing it as a statutory objective [10]. In 2022, the concept of climate-resilient cities and their legal implications were introduced. Rotterdam serves as an example of a city that has been proactive in building resilience and working towards urban sustainability in the face of climate change challenges [10]. The literature highlights the significance of climate resilience in cities such as Nantong, stressing the need for assessment, planning, construction, monitoring, and financing of climate-resilient infrastructure to mitigate the impacts of climate change and promote sustainable urban development.

It is crucial to consider factors such as urban infrastructure, economic support capacity, social development capacity, endowment of natural resources, technology adaptation capacity, and risk management capacity to enhance the resilience of cities like Nantong to climate change [11]. Governments are crucial in leading the development of climate-resilient cities and encouraging cross-regional collaboration on adaptation through forward-thinking climate planning. The exposure of cities to climate change hazards can be significantly reduced, and the overall resilience of cities to climate change can be enhanced, by financial expenditure on urban infrastructure and public facilities [5]. Network building between city actors and other cities [12], which is the foundational step in integrating urban climate resilience, is also essential for building resilience in an urban environment. For monitoring and assessing climate adaptation and development programs, it is crucial to create indicators of urban resilience [6]. Urban gardening has gained popularity in increasing the resilience of cities to climate change [13], The 100 Resilient Cities program of the Rockefeller Foundation strives to increase resilience in cities worldwide, emphasizing the importance of building resilience to the enhancement of human well-being [10]. Assessing a city's resilience to climate change means examining how well it can proactively adjust to new environmental circumstances and how quickly it can recover from the destructive effects of severe, while ensuring the well-being of its residents and preserving essential infrastructure [14, 15]. That retaining self-organizing capacity and adjusting to new conditions following extreme events is the key focus of climate-resilient thinking, which goes beyond reactive climate adaptation. Subsequently, it is of paramount importance to prioritize the aspects of government leadership, collaborative development initiatives, strategic financial investments, the promotion of urban gardening practices, the enhancement of urban infrastructure, and the formulation and implementation of resilience indicators, all of which are essential for the creation and establishment of climate-resilient urban environments such as the city of Nantong. By considering these multifaceted factors and fostering a comprehensive and interconnected approach to urban dynamics, cities can significantly augment their adaptive capacity and effectively respond to the myriad challenges posed by climate change.

In 2023, Nantong City's health system adeptly managed the transition amid the COVID-19 pandemic under municipal leadership, facilitating effective coordination of epidemic prevention and control measures. Concurrently, the city advanced significant reforms in health care, public health services, and social welfare, completing key initiatives with notable success. These efforts earned national recognition, with the city being named a model for healthy city construction for three consecutive years, alongside commendations for its comprehensive healthcare reform, which received praise during state council inspections. The city's achievements extended beyond reforms, with initiatives such as the construction of modern, patient centered hospitals and the strengthening of grassroots health governance. These advancements, along with specialized services for both the elderly and youth, garnered significant attention in the media, highlighting the city's commitment to public health and welfare. A striking example of Nantong's dedication to health and emergency services occurred when a cargo ship caught fire in the Yangtze River. With crew members injured and in need of urgent medical care, the Municipal Health Commission swiftly dispatched a helicopter to the scene. During the rescue, a patient was successfully airlifted and transported to Nantong First People's Hospital, a provincial water emergency

Fig. 1 A cargo ship in the waters of the Yangtze River. *Source* Photo taken by the Author

medical rescue base, ensuring timely treatment. This particular operation served to underscore not only the level of preparedness that the city has achieved but also the effectiveness of its emergency response mechanisms, thereby providing a compelling demonstration of the unwavering commitment displayed by local authorities as well as the health system towards ensuring the welfare and safety of the residents within the community as reflected to the photos as a kind of intervention Fig. 1.

In November 2023, Nantong City successfully executed a water emergency medical rescue drill as a culmination of continuous efforts to improve the city's emergency response skills over the previous year. Nantong improved its capacity to handle public health crises through several targeted activities, enabling prompt and coordinated responses to a variety of emergencies, including the management of 18 public health incidents and the execution of a monkeypox emergency management tabletop exercise. The city also integrated emergency medical response, toxicological disposal, epidemiological management, and psychological support into its emergency operations through thorough "sea, land, and air" rescue drills.

The formation of eight community-based medical emergency response teams, each with 20 people with mobile "backpacks" for quick deployment, was one of the major results of these efforts. By having these teams on standby, the city can better handle medical rescues, epidemic containment, and mental health crises, among other potential emergencies. This will ensure the public's safety and resilience. Nantong has achieved great progress in integrating medical defense capabilities and in strengthening emergency response infrastructure. At the forefront of this movement are the newly formed public health and medical defense synergy departments at all medical institutions at the third level and higher. Nantong further cemented its role as a leader in health governance by being the first in the province to provide a thorough framework for public health governance responsibilities for medical facilities. The formation of a cooperative partnership centered on chronic disease management was made possible by the city's hosting of a conference on the innovative development of medical-defense integration. By promoting a regionally integrated approach to healthcare, this effort set new benchmarks for the prevention and treatment of chronic diseases at the community level. Nantong is the only prefecture-level city in the province to have its tumor registration data included in the World Health Organization's International Agency for Research on Cancer (IARC) "Cancer Incidence in Five Continents" database, a testament to the city's international recognition in this field. These accomplishments highlight Nantong's dedication to city safety, innovation, and emergency response, proving the city's ability to take the lead in both international emergency response and chronic disease management at the local and international levels as shown in Fig. 2.

In 2023, Nantong City effectively navigated the smooth transition through the COVID-19 pandemic, setting a precedent in the province by introducing innovative healthcare systems. Among the key reforms were the establishment of the "general dean" system and the implementation of "three daily scheduling" protocols, which facilitated hierarchical diagnosis, classification of treatment, and multidisciplinary consultations. These measures effectively addressed the extensive outbreak earlier this year 2023, during which outpatient and emergency services managed 2.7 million COVID-19 cases. The city adopted a "one case, one program" approach, which

Fig. 2 The integration of innovation and development conference, Nantong City. *Source* Photo taken by the author

successfully treated over 4000 critically ill individuals, enabling a smooth handling of the peak in acute and critical cases.

Nantong also focused on enhancing the humanistic elements of its healthcare facilities. Initiatives such as rehabilitation gardens, sunken courtyards, mother-and-baby rooms, third bathrooms, and children's play areas contributed to creating an environment filled with warmth and care. These elements aimed to enhance the humanization of hospital environments, promoting a patient-centric care model. Building on the momentum of these reforms, the city leveraged the promotion of civilized city initiatives to launch the Beautiful Hospitals construction project, the first of its kind in the province. This initiative included 12 specific measures aimed at improving the aesthetic and functional aspects of hospitals, focusing on "environmental beauty, service beauty, humanistic beauty, and quality excellence." These efforts culminated in the recognition of several institutions in the first round of evaluations. Seven hospitals, including the First People's Hospital, the Second People's Hospital, the Sixth People's Hospital, Haimen District People's Hospital, Hai'an City People's Hospital, Shigang Central Health Centre in Tongzhou District, and Qutang Central Health Center in Hai'an City, were named as outstanding units for their contributions to the Beautiful Hospitals initiative. These accomplishments reflect Nantong's commitment to enhancing both the technical and humanistic dimensions of healthcare, positioning the city as a leader in health system reform and patient care in the province reflected in Fig. 3.

Nantong City has taken significant strides in elevating the quality of its medical services, particularly through the Beautiful Hospitals initiative, which has led to notable improvements in academic and specialist construction. These efforts have resulted in the establishment of six national key specialties, a provincial medical innovation center a provincial academician workstation, six provincial key disciplines and laboratories, and 68 provincial key clinical specialties. Additionally, 15 traditional Chinese medicine (TCM) specialties were selected by provincial TCM key specialty construction units, and the city created eight provincial-level Chinese

Fig. 3 The second people's hospital of Nantong. *Source* Photo taken by the author

medicine halls. Five hospitals were designated as provincial pilot units for high-quality public hospital development. Nantong's deepening medical reform efforts have focused on enhancing the overall healthcare system.

The city introduced the Hospital Catch-Up Plan, aimed at addressing weaknesses and reinforcing areas that needed improvement. In line with this plan, the 12 county medical communities have accelerated their transformation toward more integrated models, with Haian and Rugao becoming provincial monitoring areas for medical associations. The 2023 National Deepening Medical Reform Experience Promotion Meeting extended a warm welcome to Nantong, allowing it to demonstrate its complete medical reform achievements and acknowledging these reforms on a national level. Nantong has increased its relationship with Shanghai to boost the development and efficacy of healthcare, in addition to these internal developments. This riverine integration has resulted in the formation of agreements with 12 medical institutions in Shanghai, including 43 hospitals and 156 clinical departments. In Nantong, the medical facilities are able to take advantage of the sophisticated knowledge of more than a thousand Shanghai specialists, who travel to the city on a monthly basis for the purpose of providing consultations and performing surgical procedures. A remarkable illustration of Nantong's commitment to expanding access to high-quality healthcare is the relocation and improvement of the Chongchuan District Qinzao Street Community Health Service Centre. After more than two years of building, the facility finally began its trial operations, giving residents a more favorable environment for receiving healthcare. The refurbished facility, which is one of four regional healthcare facilities in the Chongchuan District, has improved medical conditions, a more practical layout, and increased accessibility. As a result, it is now much simpler for local residents to receive healthcare services that are both systematic and maintained in close proximity to their homes. The expansion of parking facilities and the modernization of infrastructure have made patient experience more convenient and more personally satisfying. Nantong has made significant strides toward developing a healthcare system that is not only effective but also friendly, with the goal of meeting the requirements of a population that is rich in diversity.

3 City Information Modelling in Healthcare Services

A comprehensive literature review on urban climate-resilient cities, with a focus on Nantong City, was undertaken to identify critical themes and strategies for urban resilience. This analysis encompasses governance research, environmental sustainability, urban planning, and public health measures. The study utilized a methodical approach for data collection and analysis. The methodology section provides a comprehensive examination of the research framework, data-gathering strategies, and statistical evaluation methods. The study employed a mixed-methods strategy, combining qualitative analysis of policy texts, case studies, and scholarly literature with quantitative assessments of climate resilience measurements in Nantong City. In order to evaluate papers and publications that are pertinent to the resilience of cities to climate change, the study employed a cross-sectional methodology, using Nantong City as a case study to analyze the strategy's implementation in depth.

Academic databases, government publications, and urban planning agencies were among the numerous sources employed to accumulate data. To gain a more comprehensive understanding of the triumphs and failures associated with the implementation of climate health resilience strategies, we conducted interviews with the urban planners, health officials, and policymakers of Nantong City. Descriptive statistics were utilized to evaluate public health outcomes, infrastructure development, and environmental sustainability, while correlation and regression studies examined factors influencing urban development, including green infrastructure, energy efficiency, and public health policy. The Materials section of the study provides a comprehensive list of the data sources and materials utilized, which are essential for the collection and analysis of data. This study did not employ any physical apparatus, as the research objectives were literature evaluation and data analysis. The study utilized qualitative analysis to code and analyze the textual data obtained from interviews and policy documents. Statistical methods facilitated the processing and analysis of numerical data for quantitative research. Academic journal databases (including ScienceDirect and JSTOR), Nantong City government papers, urban development blueprints, and environmental assessments were significant resources. In the Data Collection Procedure section we present the comprehensive data collection methodology, emphasizing the numerous data sources, data acquisition methodologies, and quality assurance protocols that are implemented to ensure the validity and reliability of the data. To effectively compile a comprehensive array of information, it was imperative for us to consult an extensive and diverse range of sources; this included not only various esteemed academic journals but also essential government documents, and frameworks that pertain specifically to urban development. We also conducted in-depth interviews with key and influential stakeholders who possess significant insights into the subject matter. Public data on Chinese urban resilience and climate adaptation strategies, particularly Nantong city, is available. We meticulously employed semi-structured interviews with local officials to acquire vital primary data, and concurrently, we utilized an array of scholarly and governmental publications to systematically gather secondary data that would support our research objectives.

At the conclusion of the study, we undertook a thorough and meticulous examination of each individual data set to ensure that it was indeed pertinent and aligned consistently with the primary research objectives that we had established. A multitude of validation techniques were strategically utilized to guarantee the utmost precision and dependability of the data collected throughout the research process. We used triangulation to cross-verify data from numerous sources and pay attention to outliers and irregularities during statistical studies. To confirm the accuracy of the research findings, peer evaluations were also undertaken. This method ensures that the study's findings are precise and accurately reflect the actual challenges and potential solutions that Nantong City encounters in its efforts to improve urban health and climate resilience as exemplifies in Fig. 4.

Fig. 4 Dr Zhang Fengqin (pictured with colleagues) is the lead Obstetrics and Gynaecology physician at Wolf Mountain Township Community Health Service Centre in Chongchuan District. He was honoured as a "China's Good Doctors and China's Good Nurses" monthly awardee for October 2003. *Source* Photo taken by the author

In alignment with the Nantong Three-Year Action Programme for Improving the Capacity of Primary Healthcare Services (2023–2025), the Municipal Party Committee and Municipal Government have implemented a robust framework to enhance the city's healthcare infrastructure. The Municipal Health Commission has established a dedicated leading group that is responsible for overseeing the progress of the program in order to fulfill its responsibility of ensuring effective execution. The development of a task-breakdown table, which describes precise objectives and tasks for various stakeholders participating in the healthcare system, is a critical component of this endeavor. To maintain accountability and ensure continuous improvement, the commission has instituted a monitoring and evaluation system comprising quarterly inspections, half-yearly notifications, and annual appraisals.

These evaluations aim to assess progress, identify shortcomings, and verify the achievement of triannual action plan goals. This proactive approach underscores Nantong's commitment to strengthening primary healthcare services, promoting access to quality care, and improving public health outcomes across the city. Nantong City has made remarkable progress in enhancing primary healthcare services. The proportion of primary medical institutions with one or more specialty departments has risen to 55.65% of all facilities, reflecting a significant investment in diverse healthcare capabilities that cater to the varied needs of the community. This improvement has been accompanied by the establishment of 10 new community hospitals, 67 functional centers within rural regional medical and health centers, and 97 Grade A village health centers. Additionally, 29 rural regional medical centers have been built, positioning Nantong as a leader in the province. The city has essentially realized the "15-min medical service circle," ensuring that each town has a public health center and each village has a health room. This advancement has enhanced healthcare accessibility, facilitating residents' pursuit of medical care within their localities.

The city continues to promote the integration of high-quality healthcare resources into grassroots medical institutions, reflecting a commitment to enhancing access to care across the region. Nantong City Tertiary Hospitals to Help Grassroots Medical Institutions Work on the Implementation Plan (for Trial Implementation) has launched its first phase, with 10 tertiary hospitals, including Tongda Hospital and Ruici Hospital, partnering with 20 primary healthcare institutions. These partnerships, grounded in the principle of one hospital, one policy, are focused on formulating tailored enhancement plans to strengthen local medical services. The city has also optimized its basic public health services, achieving significant recognition. Nantong's success in managing the basic public health services project led to a second-place ranking nationally, representing Jiangsu Province in the national assessment and evaluation. Furthermore, Zhang Fengqin, Chief Physician of Obstetrics and Gynaecology at the Wolshan Township Community Health Service Centre in Chongchuan District, was honored as the October 2023 "China's Good Doctor and China's Good Nurse." In recognition of its continued efforts in promoting health and well-being, Nantong was named a National Healthy City Construction Model City for the third consecutive year. This accolade, awarded by the National Ai Wei Office, is the highest honor in the field of national healthy city construction. To consolidate these achievements, the city has led the development of 10 special health actions and implemented strict systems, including the same day delivery system," "supervision system," and "regular meeting notification system. Nantong also successfully passed the 2022–2024 cycle of the National Health City Review Assessment, further solidifying its position as a leader in promoting public health and resilient urban development (See Fig. 5).

Fig. 5 A forum on optimizing the business environment and driving the development of local medicine and blood. *Source* Photo taken by the author

Nantong City has made significant strides in improving government services, particularly in healthcare and public health administration. By establishing a tripartite communication platform that connects the government, enterprises, and hospitals, the city has fostered collaboration and held symposia with pharmaceutical enterprises to enhance dialogue and cooperation. The full implementation of the Public Place Health Permit Notification and Commitment System* has streamlined processes, resulting in the issuance of 8717 onsite public place health permits. Additionally, Nantong launched a health permit expiry reminder service, notifying 1612 supervisory units of expiring permits, ensuring compliance and continuity in public health regulation.

The city's commitment to innovation in regulatory practices is reflected in its pioneering efforts in integrated supervision. Nantong led the province by implementing the "three books with the same," a comprehensive system guiding enterprises in self-regulation and credit repair, thus promoting industry autonomy and improving public health standards. Nantong City demonstrates a strong commitment to enhancing community well-being, particularly through its innovative approach to elderly care. The Elderly Disability and Dementia Prevention and Intervention Pilot Project, initiated last year 2023 exemplifies this dedication. By providing motor function training, the program aims to address the difficulties faced by families caring for elderly members with disabilities. This initiative, which has been implemented with considerable effort and dedication, has already demonstrated exceptionally impressive outcomes. This is evidenced by more than 120 participants who have reported experiencing substantial enhancements in various aspects of both their physical well-being and mental health, which are indisputably vital for the ongoing maintenance and improvement of the overall quality of health services available in the city of Nantong. The heartfelt appreciation expressed by participants and their families underscores the project's deep impact on the community. Building on this success, Nantong has become a leader in elderly healthcare. The city successfully completed the first batch of national pilot tasks for integrating medical care with elderly services and was selected as a national hospice pilot city in its third round. Among the city's 197 medical institutions, a notable 31.9% have been recognized as "elderly-friendly," ranking Nantong first in the province. The city's achievements underscore its commitment to holistic, empathetic care for the elderly, enhancing residents' quality of life Fig. 6.

By the end of 2023, Nantong City had made substantial progress in childcare services and healthcare, reflecting its ongoing commitment to social well-being and innovation. The number of childcare facilities reached 3.7 per 1000 people, and the city established 16 provincial-level model childcare institutions along with 49 provincial-level inclusive childcare institutions. Notably, Nantong became the first city in the country to launch a dedicated childcare service brand and logo, "Love Education Tongcheng." Additionally, the city took the lead in the province by developing the "Supervisory Platform for Infant and Toddler Care Services for Infants and Toddlers Under 3 Years of Age," which now covers 98% of local childcare institutions.

Fig. 6 Nantong was the first in the province to build the "Supervisory Platform for Infant and Toddler Care Services for Infants and Toddlers Under 3 Years of Age". *Source* Photo taken by the author

The city's commitment to creating a universal childcare system has been recognized by the provincial government, earning incentives through its supervisory efforts. Moreover, 98% of Nantong's childcare institutions are striving to attain the status of national birth-friendly advanced units. In terms of livelihood protection, the city efficiently completed eight provincial and three municipal projects. The Municipal Occupational Disease Prevention and Control Hospital expanded its services by opening additional outpatient clinics and independent beds.

Nantong also became the first city in the province to introduce the management of medical caregivers. Leveraging three provincial training bases, the city trained 469 medical caregivers, improving the overall standard of care for patients. Additionally, the city pioneered the "119 + 120" joint first aid model, resulting in more than 2000 successful joint rescues to date. Nantong has also prioritized the development of young medical talent. In the month of August in the year 2023, the Young Medical Professionals Cultivation Project, which is a significant undertaking by the city, achieved a noteworthy and momentous milestone by formally announcing the inaugural cohort comprising 50 exceptionally talented young medical professionals who are poised to contribute to the field. Throughout the ensuing three-year period, this comprehensive program is strategically designed to allocate substantial funding that will support a wide array of academic development initiatives, facilitate advanced training opportunities, and provide invaluable experiences for students to engage in learning endeavours outside of the geographical confines of China, all of which are underpinned by an impressive annual budget of 6.25 million yuan. The overarching objective of this innovative program is to meticulously cultivate a new generation of medical experts who are equipped with state-of-the-art training methodologies and possess robust academic credentials, thereby significantly enhancing the healthcare, technology, education, and research sectors within the city of Nantong.

Over the course of the past year, the Municipal Health and Wellness Commission of Nantong City has been diligently engaging in a variety of vital and strategic initiatives that are specifically aimed at fortifying and enhancing the overall efficacy of its healthcare system. The primary objectives of these comprehensive initiatives are centered around the recruitment and development of exceptionally qualified individuals who can meet the demands of the evolving healthcare landscape. This comprehensive strategy has successfully recruited 78 exceptional professionals from a diverse pool of applicants by distributing 129 targeted job advertisements. The commission has initiated a groundbreaking Young Medical Expert Training Program in addition to its recruitment efforts. This initiative signifies a renewed emphasis on the practical and professional development of young medical cadres, who will be indispensable to the industry's future. Through senior training courses and forums such as the "Youth Seedling Cultivation of Health Escort," this innovative initiative and others like it illustrate the city's commitment to producing the educated and trained workers required to sustain and improve its healthcare system. Nantong's unwavering dedication to the advancement of public welfare is clearly demonstrated through these multifaceted efforts, which aim to provide groundbreaking childcare services, state-of-the-art healthcare solutions, and the systematic cultivation of highly skilled medical talent. The city is actively encouraging the development of a more inclusive, health-oriented, and resilient community that can prosper in the face of modern difficulties by placing a strong emphasis on these important areas as seen in the Fig. 7.

Fig. 7 An official opening of the Shanghai University (Nantong) Geriatrics Institute/Shanghai University a senior training course for young talent in Shanghai. *Source* Photo Source taken by the author

Nantong City has also made significant advancements in the integration of information technology within its healthcare system, which has proven instrumental in enhancing healthcare delivery and infrastructure. The city's universal health information platform has been upgraded to ensure that all major medical and health units in the urban area are connected to the city health network, achieving full coverage. This digital transformation has also led to the establishment of 11 internet hospitals, broadening access to medical services and promoting telemedicine.

Nantong's efforts in digital healthcare have earned recognition, with the city's exploration of a digital medical network selected as an outstanding practice result in the 2023 Digital Jiangsu Construction* initiative. These advancements position Nantong at the forefront of healthcare innovation, leveraging technology to provide more efficient and accessible care. Further bolstering its healthcare system, Nantong has fostered academic collaborations with prestigious institutions. The opening of the Shanghai University (Nantong) Institute of Geriatrics marked the first collaboration between Nantong's medical institutions and Shanghai's leading universities. This was followed by the inauguration of the He Lin Academician Workstation at Nantong No. 4 Hospital, a significant step in enhancing the city's academic and clinical capabilities. These partnerships highlight Nantong's commitment to integrating cutting-edge research and education with its healthcare system, further advancing the quality of care available to its residents as emphasizsed in Fig. 8.

Nantong City has accelerated the construction of key healthcare infrastructure projects, reflecting its commitment to expanding and enhancing medical services. As of 2023, 48 healthcare projects are underway at city and county levels. Key initiatives include the City Hospital relocation, City Oncology Hospital complex construction, and City Emergency Hospital establishment. These projects are essential for enhancing Nantong's healthcare infrastructure in response to population growth and

Fig. 8 Inauguration of Helin workstation *Photo Source* Photo taken by the author

advancing the city's status in modern medical services. Nantong City continues to prioritize healthcare development with several key projects nearing completion, including the Tongda Hospital Nantong Medical Centre, Phase II of the Medical Complex, and the City Women's Hospital Children's Building. These projects, now in the final stages of civil interior decoration, are part of a broader effort to modernize and expand healthcare infrastructure.

In 2024, the Municipal Health and Wellness Commission will focus on advancing Nantong's healthcare sector, guided by Xi Jinping's thoughts on socialism with Chinese characteristics in the new era. Zhang Bing, Secretary of the Party Group and Director of the Municipal Health Commission, emphasized that this year's initiatives will align closely with the strategic direction set by the Municipal Party Committee and Government. With high-quality development as a central theme, the Commission will work towards ensuring that citizens feel more secure, that governance becomes more efficient, and that society is increasingly satisfied with health services. The Commission's forward-looking approach emphasizes problem-solving, goal setting, and the effective execution of strategies, all contributing to making Nantong a significant growth pole for Jiangsu Province's high-quality development. The focus will be on building a model health city, contributing to the broader goal of Chinese-style modernization, and ensuring that the health system meets both current and future needs. Through these efforts, Nantong is set to become a leader in healthcare development and modernization, making valuable contributions to both the province and the nation.

4 Discussion and Implications

The establishment of buildings that promote health, and the development of urban environments that exhibit resilience to the impacts of climate change, are of paramount importance for the advancement of urban development strategies that respond to the numerous challenges posed by the phenomenon of climate change. The Nantong City case study demonstrates the need to integrate resilience and sustainability strategies to tackle future urban challenges. As cities grow and environmental pressures mount, it becomes imperative to design buildings that prioritize both occupant well-being and sustainability. Nantong's approach to urban planning highlights the importance of adaptive infrastructure in enhancing climate resilience. Through the prioritization of health and climate resilience, the municipality enhances its ability to navigate forthcoming uncertainties while safeguarding the welfare of its inhabitants.

This case study demonstrates the global need for urban development strategies that align with sustainability goals and climate adaptation measures. Nantong's efforts show how cities can blend health-focused design with climate-resilient infrastructure to foster resilient, adaptable, and sustainable urban environments [16]. The significance of evaluating, alleviating, and adjusting to the repercussions of climate change to construct resilient urban environments has been underscored. This is congruent with the results of [17] the individuals or entities that bring to the forefront the substantial and commendable initiatives undertaken by urban areas; initiatives that are aimed at diminishing their ecological footprint while simultaneously enhancing their capacity to withstand and adapt to the multifaceted repercussions and challenges posed by climate change [18] particularly underscores the fundamentally multidisciplinary and interdisciplinary attributes that serve as the foundational elements of the research dedicated to the advancement of structures that promote health and well-being, as well as the creation of urban settings that are resilient to the challenges posed by climate change. These challenges require the cooperative efforts and collaborative engagement of a diverse array of academic disciplines and professional fields, including, but certainly not restricted to, architecture, civil engineering, and the domain of public health. In order to proficiently and comprehensively confront the diverse, multifactorial, and intricately nuanced challenges presented by the perpetually changing phenomenon of climate change, it is exceedingly crucial to implement a multidisciplinary strategy that cohesively amalgamates an array of academic fields and practical methodologies. This all-encompassing framework not only aids in the development of holistic and integrative solutions but also markedly bolsters urban resilience and simultaneously fosters sustainability within the complexities of the constructed environment.

Through the amalgamation of various domains of expertise, scholars, and professionals are granted the capacity to formulate inclusive and holistic strategies that encompass all essential dimensions of architectural design, infrastructure enhancement, and public health programs, thereby resulting in urban landscapes that are increasingly efficient, resilient, and sustainable in response to persistent transformations and challenges [19] who stressed the significance of urban resilience frameworks that consider factors such as land use and building controls [20] and [21]. The formulation of strategies aimed at fostering climate-resilient urban development is paramount in confronting the multifaceted challenges introduced by climate change. This imperative corresponds with the overarching requirement for a multidisciplinary paradigm that synthesizes perspectives from architecture, civil engineering, and public health. The Resilience Assessment Framework (RAF) measures resilience and progress.

The Comprehensive Resilience Framework (CRF) incorporates a multifaceted understanding of urban resilience, which intricately includes various dimensions such as public health, the robustness of infrastructure systems, environmental sustainability, economic vitality, social cohesion, and the effectiveness of governance structures [22]. The advancement of innovative methodologies within the realm of green building principles is of utmost importance for the effective mitigation of heat-related issues as well as for the adaptation strategies necessary to combat the multifaceted challenges posed by climate change. In the scholarly work authored by Liang (2021), there is a thorough examination of the ways in which proactive governmental leadership, coupled with collaborative efforts in urban development, significantly enhances the adaptive capacities and overall resilience of urban areas specifically designed to withstand the impacts of climate-related adversities. The integration of sustainable practices within comprehensive and robust governance frameworks is not merely beneficial but is, in fact, essential for effectively addressing the diverse and far-reaching repercussions of climate change, while simultaneously working towards the enhancement of resilience in urban environments [23].

The significance of implementing green construction methodologies in mitigating the rising temperatures and effectively responding to the challenges posed by climate change cannot be overstated, as it represents an essential component of contemporary environmental strategies. Liang (2021) meticulously examines the intricate processes that are integral to the establishment of urban environments that are resilient to climate fluctuations, while underscoring the pivotal importance of proactive government leadership coupled with collaborative development initiatives that serve to enhance adaptive capacities and bolster overall resilience in communities. To proficiently tackle the complex and multifaceted consequences of climate change while simultaneously fostering improvements in urban resilience, this comprehensive strategy emphasizes the critical necessity of harmoniously integrating sustainable construction practices within robust governance frameworks that are designed to support long-term ecological and social sustainability.

5 Summary: Strategic Measures for Building a Resilient City

The creation of resilient cities has become a significant focus in urban planning as a strategy for enhancing safety and sustainability. Research underscores the importance of urban resilience and provides recommendations for assessing and improving a city's ability to address various issues. Resilient cities exhibit indicators of rapid recovery, adaptability, and the continuance of essential services in the face of uncertainty. In order to mitigate hazards and mitigate catastrophic losses, they implement innovative urban management strategies. The City Resilience Framework (CRF) encompasses the many facets of urban resilience, including leadership, health, infrastructure, the environment, and economics.

To enhance resilience, it is imperative to consider economic geography, urbanization, and sustainability. Adapting a comprehensive strategy to the distinctive characteristics of each community is necessary for the assessment and enhancement of resilience. Innovative strategies for building resilient cities include implementing wise development practices, including resilience into governance models, and efficiently managing groundwater. The RAF serves as a metric for evaluating both resilience and progress. The CRF encompasses urban resilience, including governance, health, infrastructure, and environmental protection. The incorporation of the environment into urban design is a critical component of the development of resilient cities, necessitating a comprehensive understanding of governance, socioeconomic, and environmental aspects. This is a global imperative, and Nantong City serves as an exemplary illustration of the necessity of cultivating climate-resilient urban environments. Future initiatives should examine innovative building materials and technology while assessing their long-term social and environmental effects.

(VOCs): Volatile organic compounds.
(IPCC): The International Panel on Climate Change.
(OECD): The Organization for Economic Cooperation.
(IARC): International Agency for Research on Cancer.
(TCM): Traditional Chinese medicine.
(RAF): The Resilience Assessment Framework.
(CRF): City Resilience Framework.

References

1. Hossain I, Mahmudul Haque A, Akram Ullah S (2024) Socio-economic dimensions of climate change in urban Bangladesh: a focus on the initiatives of local governing agencies. In Climate crisis, social responses and sustainability: socio-ecological study on global perspectives, pp 293–316. Springer.
2. Huan Y, Liang T, Li H, Zhang C (2021) A systematic method for assessing progress of achieving sustainable development goals: a case study of 15 countries. Sci Total Environ 752:141875
3. Badolo M (2024) Modeling and planning interdependent critical urban infrastructures resilience to extreme events: the Badolo Cires model. Authorea Preprints.
4. Fu L, Yang X, Zhang D, Cao Y (2021) Assessment of climate-resilient city pilots in China. Chin J Urban Environ Stud 9(01):2150005
5. Cunpu L, Lishuo P, Jing, H (2023) A study of urban climate resilience and its influencing factors in China
6. Feldmeyer D, Wilden D, Kind C, Kaiser T, Goldschmidt R, Diller C, Birkmann J (2019) Indicators for monitoring urban climate change resilience and adaptation. Sustainability 11(10):2931
7. Jabareen Y (2015) The risk city. In: Springer
8. Burchard-Dziubińska M (2021) The role of urban gardening in building city resilience to climate change. Ekonomia i Środowisko.
9. Jena LP (2021) Financing climate-resilient infrastructure in cities. Emerging Econ Stud 7(2):107–114
10. Spaans M, Waterhout B (2017) Building up resilience in cities worldwide–Rotterdam as participant in the 100 resilient cities programme. Cities 61:109–116
11. Zhang H, Cheng H, Sinha A, Spötl C, Cai Y, Liu B, Kathayat G, Li H, Tian Y, Li Y (2021) Collapse of the Liangzhu and other Neolithic cultures in the lower Yangtze region in response to climate change. Sci Adv 7(48), eabi9275
12. Bilska A (2017) Embedding resilience of urban areas to climate change: a case study of Rotterdam. Urban Dev, 56 (2017)
13. Janikowska O, Jebreel AAA (2022) The effect of energy transition on the labor market. A preliminary evaluation of Poland's wind-energy industry. Polityka Energetyczna, 25
14. Schaefer D (2021) The relationship between quality of life, extreme climatological conditions and social unrest in India
15. Sharma S, Vredenburg H (1998) Proactive corporate environmental strategy and the development of competitively valuable organizational capabilities. Strateg Manag J 19(8):729–753
16. Bouramdane A-A (2024) Crafting an optimal portfolio for sustainable hydrogen production choices in Morocco. Fuel 358:130292

17. Ondrikova N, Harris JP, Douglas A, Hughes HE, Iturriza-Gomara M, Vivancos R, Elliot AJ, Cunliffe NA, Clough HE (2023) Predicting norovirus in England using existing and emerging syndromic data: Infodemiology study. J Med Internet Res 25:e37540
18. Seduikyte L, Gražulevičiūtė-Vileniškė I, Povilaitienė I, Fokaides PA, Lingė D (2023) Trends and interdisciplinarity integration in the development of the research in the fields of sustainable, healthy and digital buildings and cities. Buildings 13(7):1764
19. Tyler S, Moench M (2012) A framework for urban climate resilience. Climate Dev 4(4):311–326
20. Brummer A (2024) The data needs and solutions of a living real estate company
21. Hung C-LJ, James LA, Carbone GJ, Williams JM (2020) Impacts of combined land-use and climate change on streamflow in two nested catchments in the Southeastern United States. Ecol Eng 143:105665
22. Wang W, Wei H, Hassan H, He X (2024) Research progress and prospects of urban resilience in the perspective of climate change. Front Earth Sci 12:1247360
23. Iturriza M, Hernantes J, Labaka L (2019) Coming to action: operationalizing city resilience. Sustainability 11(11):3054

The manufacturer's authorised representative in the EU is Springer
Nature Customer Service Centre GmbH, Europaplatz 3, 69115 Heidelberg,
Germany. If you have any concerns regarding our products, please
contact ProductSafety@springernature.com

Printed and bound by CPI Group (UK) Ltd, Croydon, CR0 4YY
01/10/2026
02225936-0002